PORTFOLIO MANAGEMENT WITH HEURISTIC OPTIMIZATION

Advances in Computational Management Science

VOLUME 8

The title published in this series are listed at the end of this volume.

Portfolio Management with Heuristic Optimization

by

DIETMAR MARINGER
University of Erfurt, Germany

A C.I.P. Catalogue record for this book is available from the Library of Congress.

ISBN-10 0-387-25852-3 (HB)
ISBN-13 978-0-387-25852-2 (HB)
ISBN-10 0-387-25853-1 (e-book)
ISBN-13 978-0-387-25853-9 (e-book)

Published by Springer,
P.O. Box 17, 3300 AA Dordrecht, The Netherlands.

www.springeronline.com

Printed on acid-free paper

Preface

Managing financial portfolios is primarily concerned with finding a combination of assets that serves an investor's needs and demands the best. This includes a wide range of aspects such as the analysis of the investor's attitude towards risk, expected return and consumption; estimations of future payoffs of the financial securities and the risk associated with it have to be made; assessing the relationships between securities; determining fair prices for these securities – and finding an optimal combination of financial securities. Many of these tasks are interrelated: what is an optimal combination depends on the investor's preferences as well as on the properties of the assets, which, in return, will affect what is considered a fair price and *vice versa*.

The usual (theoretical) frameworks for portfolio management and portfolio optimization assume markets to be frictionless. Though it drives the models away from reality, this assumption has long been considered the only way to make these models approachable. However, with the advent of a new type of optimization and search techniques, *heuristic optimization*, more complex scenarios and settings can be investigated and many of these simplifying assumptions are no longer necessary.

This book is merely concerned with problems in portfolio management when there are market frictions and when there are no ready-made solutions available. For this purpose, the first two chapters present the foundations for portfolio management and new optimization techiques. In the subsequent chapters, financial models will be enhanced by problems and aspects faced in real-life such as transaction costs, indivisible assets, limits on the number of assets, alternative risk measures and descriptions of the returns' distributions, and so on. For each of these enhanced problems, a detailed presentation of the model will be followed by a description of how it can be approached with heuristic optimization. Next, the suggested approaches will be applied to empirical studies and the conclusions for financial theory will be discussed.

Non-technical Summary

The theoretical foundation to portfolio management as we know it today was laid by Harry M. Markowitz by stating a parametric optimization model. The gist of this model is to split the portfolio selection process into two steps where first the set of optimal portfolios is determined and then the investor chooses from this set that portfolio that suits her best. Markowitz's approach therefore includes (i) measuring the expected return and risk of the available assets (independently of the investor's believes and preferences), and (ii) making certain assumptions about the investor's utility functions (independently of the available assets). These two steps are then brought together in a quadratic optimization problem. This model, by now the centre of *Modern Portfolio Theory*, provoked a revised notion of risk and in due course of what is a fair risk premium.

Chapter 1 presents some aspects of the financial theory underlying this contribution, including the portfolio selection problem in a Markowitz framework and selected related and follow-up literature. In addition, two equilibrium models will be presented: the *Capital Asset Pricing Model (CAPM)*, which, in this contribution, will be used to generate data for equilibrium markets, and the concurring *Arbitrage Pricing Theory (APT)* for which relevant risk factors will be identified. The chapter concludes after a short presentation of alternative approaches to portfolio management.

With all its merits, the Markowitz model has a major downside: to get a grip of the computational complexity, it has to rely on a number of rather strict technical assumptions which are more or less far from reality: markets are assumed to be perfect in the sense that there are neither taxes nor transactions costs and assets are infinitely divisible; investors make their decisions at exactly one point in time for a single-period horizon; and the means, standard deviations and correlation coefficients are sufficient to describe the assets' returns. Though there exists no closed-form solution for the Markowitz model, the simplifying assumptions allow for a solution with standard software in reasonable time if the number of assets is not too large.

The limitations of the original Markowitz framework have stimulated a number of extended or modified models. These models allow for valuable insights – yet still have to make simplifying assumptions in order to be solvable: seemingly simple questions such as adding proportional plus minimum transactions costs, taking into account that usually stocks can be traded in whole-numbered lots, or allowing for non-parametric empirical distributed returns are unsolvable with standard methods. It therefore appears desirable to have alternative methods that can handle highly demanding optimization problems.

One way out of this dilemma is *heuristic optimization* (*HO*). The techniques employed in HO are mostly general purpose search methods that do not derive the solution analytically but by iteratively searching and testing improved or modified solutions until some convergence criterion is met. Since they usually outperform traditional numerical procedures, they are well suited for empirical and computational studies. **Chapter 2** presents some general concepts and standard HO algorithms.

Having introduced some basic concepts, heuristic optimization techniques are applied to some portfolio selection problems which cannot be solved with other, more traditional methods.

The effects of magnitude of initial wealth, type of transactions costs as well as integer constraints on the portfolio selection problem will be discussed based on DAX data in **chapter 3**. We distinguish a number of cases where investors with different initial wealth face proportional costs and/or fixed transactions costs. As the associated optimization problem cannot be solved with standard optimization techniques, the literature so far has confined itself to rather simple cases; to our knowledge, there are no results for an equally comprehensive model. This problem is usually approached by first solving the problem without these aspects and then fitting the results on the real-world situation. The findings from the empirical study illustrate that this might lead to severely inferior solutions and wrong decisions: Unlike predicted by theory when the usual simplifications apply, investors are sometimes well-advised to have a rather small number of different assets in their portfolios, and the optimal weights are not directly derivable from those for frictionless markets.

For various reasons, investors tend to hold a rather small number of different assets in their portfolios. Also, it is a well-known fact that much of a portfolio's diversification can be achieved with a rather small number of assets – yet, to our knowledge there exist only rough estimates based on standard rules or simple simulations to evaluate this fact. **Chapter 4** focuses on the selection problem under cardinality constraints, i.e., when there are explicit bounds on the number of different assets. The empirical study uses data for the DAX, FTSE and S&P 100. The main results are that small (yet well-selected) portfolios can be almost as well-diversified as large portfolios and that standard rules applied in practice can be outperformed.

Chapter 5, too, investigates the effects of cardinality constraints yet in a different setting where not just one specific portfolio, but the whole so-called "efficient sets" are to be identified. In order to meet the high computational complexity of this problem, a new algorithm is developed and tested against alternative optimization heuristics. With the focus on the computational aspects, it is shown that hybrid algorithms, combining aspects from different heuristic methods can be superior to basic algorithms and that heuristic optimization algorithms can be modified according to particular aspects in the problems. With this new algorithm at hand, the highly demanding optimization problem can now be approached.

The usual definition of "financial risk" captures the assumed (positive and negative) deviations from the expected returns. In some circumstances, however, the investor might be more interested in the maximum loss with a certain probability or the expected loss in the worst cases. Hence, alternative risk measures such as *Value at Risk* (*VaR*) and *Expected Shortfall* (*ES*) have gained considerable attention. **Chapter 6** is concerned with the question of whether these new risk measures actually make good risk constraints when the investor is interested in limiting the portfolio's potential losses. Based on empirical studies for bond markets and stock markets, we find that VaR has severe shortcomings when it is used as an explicit risk constraint, in particular when the normality assumption of the expected returns is abandoned (as has often been demanded by theory and practice).

The *Arbitrage Pricing Theory* (*APT*) is sometimes considered superior to other equilibrium pricing models such as, e.g., the CAPM as it does not use an (actually unobservable) market portfolio but a set of freely selectable (and observable) factors. The major shortfall of the APT, however, is that there are no straightforward or *a priori* rules of how to find the ideal set of factors: there are not always "natural" candidates for factors, standard choices do not work equally well for all assets (or are not applicable for other reasons). Given a set of potential candidates, the associated selection problem is computationally extremely demanding. **Chapter 7** finds that this model selection problem, too, can be approached with heuristic search methods. The selected combinations of factors are likely to identify fundamentally plausible indicators and they are likely to explain a considerable share of the variation in the assets' returns.

Chapter 8 concludes and presents an outlook on possible follow-up research.

The prime focus of this contribution is on individual investment decisions under market frictions. The main part of the study will therefore consider individual investors who already have estimates for future returns and risks but face the problem of how to translate these estimates into optimal portfolio selections (chapters 3 – 6) or how to translate estimates for aggregated market factors into pricing models for individual assets (chapter 7) in the first place. In all of these problems, the investors are considered to be rational and risk averse price takers operating in equilibrium markets.

This contribution therefore aims to answer financial management problems that are well identified in financial theory and faced by the investment industry but could not yet be answered satisfactorily by the literature. Due to the restrictions in traditional optimization methods, the respective models had to rely on simplifying assumptions and stylized facts that restrained the applicability of the results. In this contribution, an alternative route is chosen and new optimization methods are applied that are capable of dealing with otherwise unanswerable problems. The results show that this approach is capable of identifying shortcomings of traditional approaches, that market frictions and complex constraints can now easily and completely be incorporated in the optimization process without the usual prior simplifications (which, as will be shown, can even be misleading), and that problems can be solved for which just approximations or rules of the thumb existed so far.

The results from these studies also indicate the gain from the application of new methods such as heuristic optimization: Models and problems can be investigated that allow for more complexity and are therefore closer to reality than those approachable with traditional methods, which eventually also contributes to a better understanding of financial markets.

Acknowledgement

This project would not have been possible without the input, suggestions, participation, guidance, moral support, and comments of many people. My special thanks go to Edwin O. Fischer and Peter Winker for stimulating, fostering, and supporting my interest in finance and computational economics; for guiding, encouraging and supporting my first steps in the world of academia; for valuable comments, supervision and critique; and, not least, for their patience. Also, I would like to express my gratitude to W. Burr, U. Derigs, E. Dockner, M. Gilli, H.-O. Günther, R. Hartl, C. and U. Keber, H. Kellerer, E. Kontoghiorghes, L. Kruschwitz, K. Krycha, A. Lehar, M. Leitinger, K. Lerch, E. Loitlsberger, G. Meeks, U. Ott, V. Priemer, B. Rustem, C. Strauss, S. Sundaresan, F. W. Wagner, U. Wagner, P. Walgenbach, J. Zechner, my colleagues (past and present), to journal editors and referees, numerous anonymous participants at conferences and research seminars, the "Habilitation" committee at the Faculty of Economics, Law and Social Sciences at the University of Erfurt who accepted this book for a *habilitation thesis*, C. van Herwaarden, H. Drees, the series editors from Springer – and many, many more.

Contents

Chapter 1

Portfolio Management

1.1 Portfolio Optimization

1.1.1 Mean-Variance Analysis

A common property of investment opportunities is that their actual returns might differ from what has been expected; or in short: they are *risky*. This notion of financial risk, defined by the (potential) deviation from the expected outcome, includes not only a lower than expected outcome (*downside risk*) but also that the actual return is better than initially expected (*upside risk*) because of positive surprises or non-occurrences of apprehended negative events.[1]

When all available information and expectations on future prices are contained in current prices,[2] then the future payoffs and returns can be regarded and treated as random numbers. In the simplest case, the returns of an asset i can be described with the normal distribution: the *expected value (mean)* of the returns, $E(r_i)$, and their *variance*, σ_i^2, (or its square root, σ_i, in the finance literature usually referred to as *volatility*) capture all the information about the expected outcome and the likelihood and range of deviations from it.

---

[1] See also section 1.1.3, where alternative concepts of risk and risk measures as well as sources of risk will be presented.

[2] Fama (1970) has initiated a still ongoing discussion about *information efficiency*. Note that in portfolio management, these expectations are not necessarily rational but can also reflect herding behavior, market anomalies and the like. See also section 1.3.

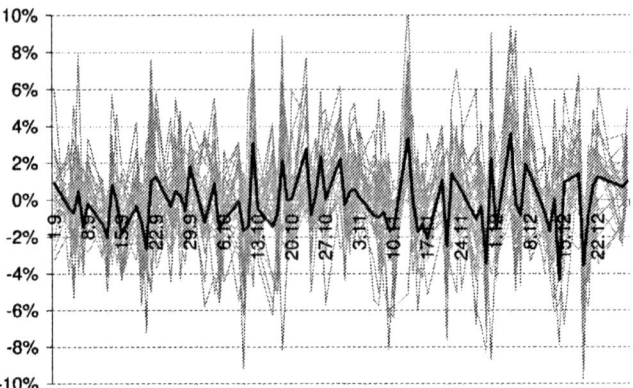

Fig. 1.1: Daily returns of the DAX (black line) and the stocks contained in it (gray lines) for the 4th quarter of the year 2000

When comparing investment opportunities and combining them into portfolios, another important aspect is how strong their returns are "linked", i.e., whether positive deviations in the one asset tend to come with positive or negative deviations in the other assets or whether they are independent. If the assets are not perfectly positively correlated, then there will be situations where one asset's return will be above and another asset's return below expectance. Hence, positive and negative deviations from the respective expected values will tend to partly offset each other. As a result, the risk of the combination of assets, the *portfolio*, is lower than the weighted average of the risks of the individual assets. This effect will be the more distinct the more diverse the assets are. The intuition is that similar firms (and hence their stocks) do similarly poorly at the same time whereas in heterogeneous stocks, some will do better than expected while others do worse than expected. The positive and negative deviations from the expected values will then (to some degree) balance, and the actual deviation from the portfolio's expected return will be smaller than would be the deviation from an asset's expected return even when both have the same expected return. Figure 1.1 illustrates this effect for the daily returns of the German DAX and the 30 stocks included in it: the index's daily returns appear more stable over time (i.e., exhibit less risk) than the assets' included in the index, yet the reduction in risk does not necessarily come with a reduction in the average return.

Technically speaking, the risk and return of a portfolio $\mathcal{P}$ consisting of N risky assets can be treated as a convolution of the individual assets' returns and covariances when the included assets can be described by the distributions of their returns. The portfolio $\mathcal{P}$ will then have an expected return

$$E(r_\mathcal{P}) = \sum_{i=1}^{N} x_i \cdot E(r_i) \tag{1.1}$$

and a variance

$$\sigma_\mathcal{P}^2 = \sum_{i=1}^{N}\sum_{j=1}^{N} x_i \cdot x_j \cdot \sigma_{ij} \tag{1.2}$$

where x_i is the share of asset i in the portfolio, hence $\sum_{i\in\mathcal{P}} x_i = 1$. σ_{ij} denotes the covariance between the returns of i and j return with $\sigma_{ii} = \sigma_i^2$ and $\sigma_{ij} = \sigma_{ji} = \sigma_i \cdot \sigma_j \cdot \rho_{ij}$ where $\rho_{ij} \in [-1, 1]$ is the *correlation coefficient*.

Assuming that all assets in the portfolio have the same weight $x_i = x = 1/N$, then equations (1.1) and (1.2) can be rewritten as

$$E(r_\mathcal{P}) = \sum_{i=1}^{N} \frac{1}{N} \cdot E(r_i)$$

$$= \bar{r}_i$$

and

$$\sigma_\mathcal{P}^2 = \sum_{i=1}^{N}\sum_{j=1}^{N} \frac{1}{N} \cdot \frac{1}{N} \cdot \sigma_{ij}$$

$$= \frac{1}{N} \cdot \underbrace{\frac{1}{N} \cdot \sum_{i=1}^{N} \sigma_i^2}_{=\bar{\sigma}_i^2} + \frac{N-1}{N} \cdot \underbrace{\frac{1}{N\cdot(N-1)} \cdot \sum_{i=1}^{N}\sum_{\substack{j=1 \\ j\neq i}}^{N} \sigma_{ij}}_{=\bar{\sigma}_{ij}}$$

$$= \bar{\sigma}_{ij} + \frac{1}{N} \cdot (\bar{\sigma}_i^2 - \bar{\sigma}_{ij})$$

where $\bar{r}_i$ is the average of the expected returns of the N assets, $\bar{\sigma}_i^2$ represents the average of the N variances and $\bar{\sigma}_{ij}$ is the average of the $N\cdot(N-1)$ covariances between assets $i \neq j$. The larger N, the closer the portfolio's variance will be to the average covariance while the expected value for the return remains the same. The diversification effect will be the larger the lower the correlation between the assets'

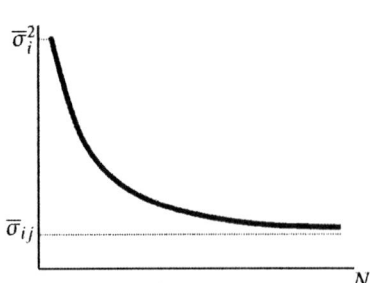

Country	$1 - \overline{\sigma}_{ij}/\overline{\sigma}_i^2$
Switzerland	56.0%
Germany	56.2%
Italy	60.0%
U.K.	65.5%
France	67.3%
United States	73.0%
Netherlands	76.1%
Belgium	80.0%
International stocks	89.3%

Fig. 1.2: Average diversification under $\overline{\sigma}_i^2$ and $\overline{\sigma}_{ij}$ with equal asset weights

Tab. 1.1: Percentage of risk that can be diversified when holding a portfolio rather than a single "average" stock (values based on Solnik (1973) and Elton, Gruber, Brown, and Goetzmann (2003))

returns: $1 - \overline{\sigma}_{ij}/\overline{\sigma}_i^2$ is the ratio of risk that can be diversified when holding an equally weighted portfolio rather than a single "average" stock. The less the stocks in the market are correlated the more risk can be eliminated and the more advantageous is the portfolio over an investment into a single stock. Figure 1.2 depicts this effect on average when N assets are randomly picked and are given equal weights. Table 1.1 summarizes for some major stock markets what fraction of the average stock's variance could be avoided by holding an equally weighted portfolio. As can be seen, the diversification is the highest when diversification is internationally rather than nationally; an effect that holds even when currency exchange risks are considered[3] and, under a considerably more sophisticated methodology, when market declines are contagious[4].

The effect of diversification can be exploited even more when the weights x_i are optimized for each asset. Assuming the simple case of a two asset market, the investor can split her endowment into a fraction x_1 for asset 1 and $x_2 = 1 - x_1$ for asset 2. The expected return and risk of the resulting portfolio can then be rewritten

---

[3] See Solnik (1974).

[4] See de Santis and Gérard (1997).

as

$$E(r_P) = x_1 \cdot E(r_1) + (1 - x_1) \cdot E(r_2) \tag{1.3}$$

and

$$
\begin{aligned}
\sigma_P^2 &= x_1^2 \cdot \sigma_1^2 + (1 - x_1)^2 \cdot \sigma_2^2 + 2 \cdot x_1 \cdot (1 - x_1) \cdot \sigma_{12} \\
&= x_1^2 \cdot \left(\sigma_1^2 + \sigma_2^2 - 2 \cdot \sigma_{12}\right) - 2 \cdot x_1 \cdot \left(\sigma_2^2 - \sigma_{12}\right) + \sigma_2^2.
\end{aligned}
\tag{1.4}
$$

Solving (1.3) for x_1 yields

$$x_1 = \frac{E(r_P) - E(r_2)}{E(r_1) - E(r_2)}. \tag{1.5}$$

Substituting (1.5) into (1.4) shows that the portfolio's variance is a parabolic function of the portfolio's expected return.

With every additionally available asset there is a chance of further risk diversification, and the left-hand side border of the opportunity set will be pushed even farther to the left. However, it is not just the number of different assets in the portfolio[5] but merely their correlation that contributes to the diversification. In the words of Markowitz (1952, p. 89): "Not only does [portfolio analysis] imply diversification, it implies the 'right kind' of diversification for the 'right reason.'" This is illustrated in Figure 1.3: the lower the correlation, the more the curve is bent, i.e., the higher the curvature of the opportunity set. When there are more than two assets in the portfolio, the opportunity set is no longer represented by a line but by a whole area: Any point within the area in Figure 1.4 represents a feasible portfolio for a given combination of risk and return which can be selected by computing the respective weights, x_i.

Harry M. Markowitz was the first to come up with a parametric optimization model to this problem which meanwhile has become the foundation for *Modern Portfolio Theory* (*MPT*).

---

[5] Chapter 4 will focus on this aspect in more detail.

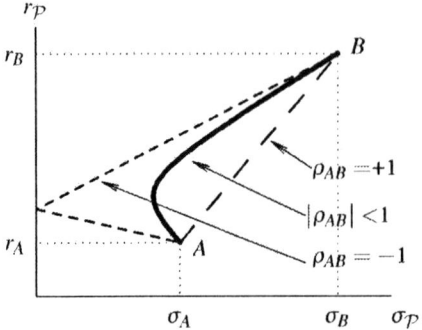

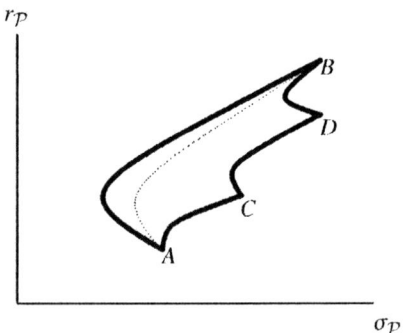

Fig. 1.3: Opportunity set for a two asset
portfolio depending on ρ_{AB}

Fig. 1.4: Opportunity set for a four asset
portfolio (dotted line: Opportunity set with
A and B only)

1.1.2 Modern Portfolio Theory (MPT)

1.1.2.1 The Markowitz Model

In his seminal paper,[6] Markowitz (1952) considers rational investors who want to
maximize the expected utility of their terminal wealth at time T, $E(\mathcal{U}(w_T))$. Investors are price takers and make their sole investment decision at time 0. If an
investor prefers more terminal wealth to less and is risk averse, then her utility function $\mathcal{U}$ with respect to terminal wealth w_T has the properties

$$\frac{\partial \mathcal{U}}{\partial w_T} > 0 \quad \text{and} \quad \frac{\partial^2 \mathcal{U}}{\partial w_T^2} < 0.$$

If the expected terminal wealth is $w_T = w_0 \cdot (1 + r_P)$, where w_0 is the (known) initial
wealth and r_P is the (risky) return of the investment over the single-period horizon
$[0, T]$ and if the investor's utility is quadratic of the type

$$\begin{aligned}
\mathcal{U}(w_T) &= \beta \cdot w_T - \gamma \cdot w_T^2 \\
&= \beta \cdot w_0 \cdot (1 + r_P) - \gamma \cdot w_0^2 \cdot (1 + r_P)^2 \\
&= \underbrace{(\beta \cdot w_0 - \gamma \cdot w_0^2)}_{\equiv a} + \underbrace{(\beta \cdot w_0 - 2 \cdot \gamma \cdot w_0^2)}_{\equiv b} \cdot r_P - \underbrace{(\gamma \cdot w_0^2)}_{\equiv c} \cdot r_P^2,
\end{aligned}$$

---

[6] For a praise of the work of Markowitz, see, e.g., Varian (1993).

where the only risky element is r_P, then her expected utility is

$$E(\mathcal{U}) = E\left(a + b \cdot r_P - c \cdot r_P^2\right)$$
$$= a + b \cdot E(r_P) - c \cdot E\left(r_P^2\right).$$

By the relationship $E\left(r_P^2\right) = \sigma_P^2 + E(r_P)^2$, the expected utility can be rewritten as

$$E(\mathcal{U}(w_T)) = a + E(r_P) \cdot (b + c \cdot E(r_P)) - c \cdot \sigma_P^2. \tag{1.6}$$

This implies that the expected returns and (co-)variances contain all the necessary information not only when the returns are normally distributed (and, hence, are perfectly described with mean and variance), but also for arbitrary distributions when the investor has a quadratic utility function. More generally, it can be shown that the mean-variance framework is approximately exact for any utility function that captures the aspects *non-satiation* and *risk aversion*.[7] Some subsequent models also assume decreasing absolute risk aversion, i.e., $\partial \mathcal{A}(w_T)/\partial w_T < 0$ with $\mathcal{A}(w_T) = \mathcal{U}'(w_T)/\mathcal{U}''(w_T)$, a property captured, e.g., by a logarithmic utility function, $\mathcal{U}(w_T) = \ln(w_T)$.[8]

The Markowitz model also assumes a perfect market without taxes or transaction costs where short sales are disallowed, but securities are infinitely divisible and can therefore be traded in any (non-negative) fraction.

Given this framework, the identification of the optimal portfolio structure can be defined as the quadratic optimization problem[9] of finding the weights x_i that assure the least portfolio risk σ_P^2 for an expected portfolio return of $r_P = r^*$.[10] The Markowitz portfolio selection model therefore reads as follows:

$$\min_{x_i} \sigma_P^2 \tag{1.7a}$$

---

[7] See, e.g., Alexander and Francis (1986, chapters 2 and 3) or Huang and Litzenberger (1988, chapters 1–2).

[8] See, e.g., Elton, Gruber, Brown, and Goetzmann (2003, chapter 10).

[9] See sections 2.1.2.3 and 2.1.2.4.

[10] For better legibility, the expectance operator $E(\cdot)$ will be dropped henceforth.

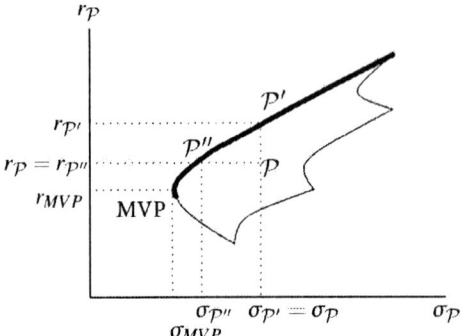
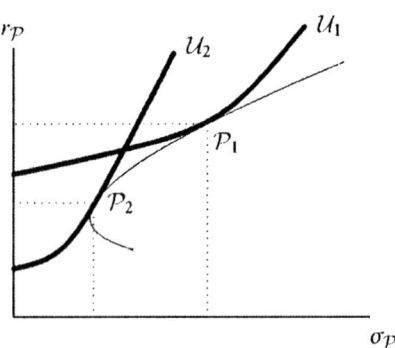

Fig. 1.5: Opportunity set and efficient line in a Markowitz framework

Fig. 1.6: Portfolio selection in a Markowitz framework depending on the investor's attitude towards risk

subject to

$$\sigma_P^2 = \sum_i \sum_j x_i \cdot x_j \cdot \sigma_{ij} \tag{1.7b}$$

$$r_P = r^* \tag{1.7c}$$

$$r_P = \sum_i x_i \cdot r_i \tag{1.7d}$$

$$\sum_i x_i = 1. \tag{1.7e}$$

$$x_i \in \mathbb{R}_0^+ \ \forall i. \tag{1.7f}$$

This optimization problem has a solution when the following technical conditions hold:

$$\min_i r_i \leq r_P \leq \max_i r_i$$

$$\sigma_i > 0 \ \forall i$$

$$\rho_{ij} > -1 \forall (i, j)$$

$$\exists (i \neq j) \text{ such that } r_i \neq r_j.$$

Depending on the covariance matrix, the portfolio with the lowest expected return is not necessarily the portfolio with the least risk. In this case, the *Minimum*

Variance Portfolio (MVP) has the least risk, as can be seen from Figure 1.5. Hence, searching for the portfolio structure with least risk for the portfolio's expected return of r^* is reasonably only for $r^* \geq r_{MVP}$. It is also apparent that rational investors will choose portfolios on the "upper" frontier of the opportunity set, represented with a bold line: For any portfolio $\mathcal{P}$ that is not on the border, there exists a portfolio $\mathcal{P}''$ with the same expected return but less risk, and a portfolio $\mathcal{P}'$ with equal risk but higher expected return. In this case $\mathcal{P}$ is an *inferior portfolio* whereas $\mathcal{P}'$ and $\mathcal{P}''$ are both *efficient portfolios*. The upper bound of the opportunity set above the MVP is therefore called *efficient set* or *efficient line*.

In model (1.7), the target expected return r^* is chosen exogenously and might therefore well be below r_{MVP} since the MVP is not known beforehand. This pitfall can be avoided by combining the return constraint (1.7c) and the corresponding risk. The original objective function (1.7a) of minimizing the risk can then be replaced with maximizing the expected return, diminished by the incurred risk,[11]

$$\max_{x_i} \left(\lambda \cdot r_{\mathcal{P}} - (1 - \lambda) \cdot \sigma_{\mathcal{P}}^2 \right), \tag{1.7a*}$$

where the trade-off between risk and return is reflected. The efficient line can then be identified by solving this problem for different, exogenously determined values of $\lambda \in [0, 1]$: If $\lambda = 1$, the model will search for the portfolio with the highest possible return regardless of the variance. Lower values for λ put more emphasis on the portfolio's risk and less on its expected return. With $\lambda = 0$, the MVP will be identified.

Due to the convexity of the Markowitz efficient line, the marginal risk premium is decreasing when there are only risky assets; also, an efficient portfolio's variance can be expressed as a function of its expected return. The portfolio selection process therefore includes the determination of the set of efficient portfolios and the identification of the optimal portfolio where the marginal risk premium equals the marginal utility (see Figure 1.6). It is one of the major contributions of the Markowitz models that the efficient set can be determined without actually knowing the investor's exact utility: The efficient set can be determined without explicit knowledge of the investors attitude towards risk (as long as investors are rational and risk averse).

---

[11] Note also the correspondence to equation (1.6).

The non-negativity constraint not only inhibits an analytic solution, it also makes the standard Markowitz model NP-hard.[12] Nonetheless, for a reasonably small number of different assets, N, it can be solved numerically with standard optimization software for *quadratic optimization problems*[13] within reasonable time.[14]

1.1.2.2 The Black Model

According to constraint (1.7f), any asset's weight must be a non-negative real number. If the non-negativity constraint is removed from the original set of assumptions and replaced with

$$x_i \in \mathbb{R} \ \forall i, \tag{1.7f*}$$

i.e., any asset's weight can be any real number – as long as constraint (1.7e) is met and they add up to 1. Negative asset weights represent *short sales* where the investor receives today's asset price and has to pay the then current price in future.

This modification is done by Black (1972), who is therefore able to find an exact analytic solution for this simplified portfolio selection problem. With short sales allowed, a closed-form solution exists and the efficient portfolio's risk and asset weights for a given level of return, $r_P = r^*$, can be determined by

$$\sigma_P^2 = \begin{bmatrix} r_P & 1 \end{bmatrix} A^{-1} \begin{bmatrix} r_P \\ 1 \end{bmatrix} = \frac{a - 2 \cdot r_P + c \cdot (r_P)^2}{a \cdot c - b^2} \tag{1.8a}$$

with

$$A = \begin{bmatrix} a & b \\ b & c \end{bmatrix} = \begin{bmatrix} r' \\ I' \end{bmatrix} \Sigma^{-1} \begin{bmatrix} r & I \end{bmatrix} \tag{1.8b}$$

and

$$x = \Sigma^{-1} \begin{bmatrix} r & I \end{bmatrix} \begin{bmatrix} a & b \\ b & c \end{bmatrix}^{-1} \begin{bmatrix} r_P \\ 1 \end{bmatrix} \tag{1.8c}$$

---

[12] See Garey and Johnson (1979). A classification of computational complexity, including the group of the most demanding, namely NP complete, problems, will follow in section 2.1.2.1.

[13] See sections 2.1.2.3 and 2.1.2.4.

[14] Jagannathan and Ma (2003) present a way of incorporating the non-negativity constraint by modifying the covariance matrix – which, however, does not lower the computational complexity of the problem.

where $r = [r_i]_{N \times 1}$ is the vector of the securities' expected returns, $\Sigma = \left[\sigma_{ij} \right]_{N \times N}$ is the covariance matrix, and I is the unity vector.[15] The return and risk of the Minimum Variance Portfolio are b/c and $1/c$, respectively.

The Black Model is sometimes used as a benchmark for efficient portfolios under additional constraints, merely due to its closed form solution, but also because it comes with some convenient properties. One of these is the fact that a linear combination of two efficient portfolios is again efficient – and *vice versa*: Any efficient portfolio can be represented as a linear combination of any two other efficient portfolios. Hence, knowing just two portfolios on the efficient line allows the replication of any efficient portfolio. On the other hand, it might produce solutions that cannot be translated into practice: when short sales are permitted in the model, the solution is likely to assign negative weights to at least some of the assets. Ignoring these results by setting the respective weights to zero and readjusting the remaining weights so that they add up to one again might result in inefficient solutions.[16]

Dyl (1975) acknowledges that the assumption of unlimited short sales might be far from reality. He therefore introduces limits for short sales, $x_i \geq x^\ell$ with $x^\ell \leq 0$ – which also leads to implicit upper limits since the available resources become restricted.

1.1.2.3 The Tobin Model

Tobin (1958, 1965) removes the condition that all assets must be risky. If the endowment is invested into a safe asset s and some risky portfolio $\mathcal{T}$, then the return of the resulting portfolio $\mathcal{P}$ is

$$r_{\mathcal{P}} = \alpha \cdot r_s + (1 - \alpha) \cdot r_{\mathcal{P}}. \tag{1.9}$$

By assumption, s is risk-free, and therefore $\sigma_s^2 = \sigma_{s\mathcal{T}} = 0$; hence, the portfolio's volatility is simply $\sigma_{\mathcal{P}} = \alpha \cdot \sigma_{\mathcal{T}}$:

$$\sigma_{\mathcal{P}}^2 = \alpha^2 \cdot \underbrace{\sigma_s^2}_{=0} + (1 - \alpha)^2 \cdot \sigma_{\mathcal{T}}^2 + \underbrace{2 \cdot \alpha \cdot (1 - \alpha) \cdot \underbrace{\sigma_{s\mathcal{T}}}_{=0}}_{=0}$$

---

[15] See also Roll (1977, p. 160).

[16] See, e.g., the results in chapter 4 as well as Sharpe (1991).

$$\Rightarrow \sigma_P = (1 - \alpha) \cdot \sigma_T. \tag{1.10}$$

Solving (1.10) for α and substituting the result into (1.9) reveals the linear relationship between the portfolio's return and volatility in a Tobin framework:

$$\alpha = 1 - \frac{\sigma_P}{\sigma_T}$$

$$r_P = r_s + (r_T - r_s) \cdot \frac{\sigma_P}{\sigma_T}. \tag{1.11}$$

This result has far reaching implications: Unlike in the Markowitz model, the efficient line is no longer a curve but a straight line in the mean-volatility space (see Figure 1.7), and any portfolio on this Tobin efficient line is a combination of the safe asset and the portfolio T: As can be seen for either utility curve in Figure 1.8, both the investor with high and the investor with low risk aversion will be better off when the safe asset is part of the overall portfolio than she would be without it, i.e., if she would be restricted to a Markowitz efficient portfolio with risky assets only.[17] As a consequence, the different levels of risk aversion and different utility curves will lead to different values for α but not to different structures within the tangency portfolio T. Hence, the optimization of the portfolio T can be separated from the individual investment decision (*separation theorem*).

With the linear relationship between risk and return, the marginal risk premium is constant. Finding the optimal structure for the portfolio T therefore is equivalent to finding the asset weights that maximize the risk premium per unit risk, i.e., the slope of the Tobin efficient line, θ_T, with

$$\theta_T = \frac{r_T - r_s}{\sigma_T}. \tag{1.12}$$

Thus, the Tobin efficient line in equation (1.11) can be rewritten as

$$r_P = r_s + \theta_T \cdot \sigma_P, \tag{1.11*}$$

going through $(0; r_s)$ and $(\sigma_T; r_T)$. Graphically speaking, it should be tangent to the efficient line of the "risky assets only" portfolios; T is therefore referred to as *tangency portfolio*. The combination with any other portfolio T' would result in a lower risk premium, i.e., $\theta_{T'} < \theta_T \ \forall T' \neq T$. When short sales are not allowed and

---

[17] If short sales are not allowed, a necessary technical condition is $r_T > r_s$.

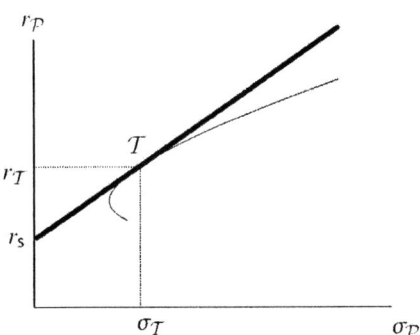
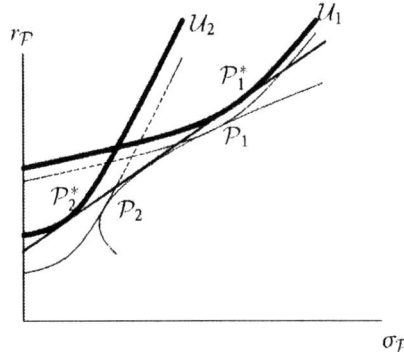

Fig. 1.7: Efficient Line for the Tobin model (curved thin line: efficient "risky assets only" portfolios according to Markowitz)

Fig. 1.8: Portfolio selection in a Tobin framework (thick lines: utility curves tangent to the Tobin efficient line; thin dashed lines: utility curves tangent to the Markowitz efficient line)

all the assets' weights must be non-negative, T is one particular portfolio from the Markowitz efficient set where, again, there exists no closed-form solution. If risky assets may not be short sold, the investor should consider a combination of safe asset plus the original Markowitz model, whenever the condition $r_s < \max_i\{r_i\}$ holds.

When there are no restrictions on short sales, then T will be located on the respective efficient line from the Black model; this framework is usually referred to as *modified Tobin model*. The exact solution for T's asset weights, return and risk, respectively, can be computed according to

$$x_T = \Sigma^{-1} \begin{bmatrix} r & I \end{bmatrix} \begin{bmatrix} \frac{1}{b-r_s\cdot c} \\ \frac{-r_s}{b-r_s\cdot c} \end{bmatrix} \tag{1.13a}$$

$$r_T = \frac{a - r_s \cdot b}{b - r_s \cdot c}$$

$$\sigma_T^2 = \frac{a - 2 \cdot b \cdot r_s + c \cdot r_s^2}{(b - c \cdot r_s)^2} \tag{1.13b}$$

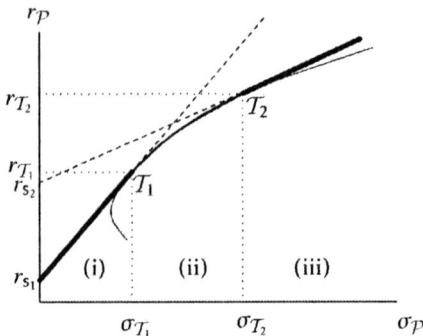

Fig. 1.9: Efficient line in a Brennan framework

with the parameters a, b, and c as introduced in (1.8b). Substituting and rearranging then implies that the optimal slope will be

$$\theta_T = \sqrt{a - 2 \cdot r_s \cdot b + r_s^2 \cdot c}.$$

Note that the combination of the safe asset plus Black model (i.e., no constraints on short selling) will lead to a reasonable solution only if $r_s < b/c$, i.e., if the safe return is lower than the return of the Minimum Variance Portfolio.

When investors want to have a portfolio with an expected return exceeding T's, $r_p > r_T$, then this implies $\alpha < 0$, i.e., going short in the safe asset, as is the case for the investor with utility curve $\mathcal{U}_1$ in Figure 1.8. Brennan (1971) distinguishes different risk-free rates for lending, r_{s_1}, and borrowing, r_{s_2}, where usually $r_{s_1} < r_{s_2}$. In his model, the investor ends up with two tangency portfolios leading to three different strategies: (i) investing in the safe asset and the respective tangency portfolio T_1 for $r_p \in [r_{s_1}, r_{T_1}]$; (ii) investing along the Markowitz efficient line (without safe asset) for $r_p \in [r_{T_1}, r_{T_2}]$; and (iii) going short in the safe asset (i.e., taking a loan at r_{s_2}) and investing the additional amount together with the initial endowment into portfolio T_2 for $r_p > r_{T_2}$. Figure 1.9 illustrates this result: the left and right bold lines represent the solutions for (i) and (iii), respectively, and the solutions according to strategy (ii) are represented by the medium bold curve connecting T_1 and T_2. Note that in the absence of transactions costs and perfect markets with $r_{s_1} = r_{s_2} = r_s$, the Brennan model collapses into the original Tobin model and $T_1 = T_2 = T$.

1.1.2.4 Further Advances

It is a well-known fact that the empirical distributions of asset returns frequently exhibit skewness and fat-tails (i.e., third and fourth moments that differ from the normal distribution's) or higher moments are not even existing.[18] The normality assumption (as a theoretically necessary prerequisite for the Markowitz model when the utility functions are not quadratic) has often been seen as unrealistic. Early attempts[19] include preferences for higher moments than the variance in their analyses.[20] These extensions, however, lead to severe problems in terms of solvability and complexity of the resulting models. A more usual way is to transform the data in a way that they are closer to the normal distribution. The simplest (and by now standard) transformation would be to use first differences of the logarithms of the prices (i.e., *continuous returns*), $r^c = \ln\left(S_t/S_{t-1}\right) = \ln(S_t) - \ln(S_{t-1})$, rather than the traditional "change : price" ratio (i.e., *discrete returns*), $r^d = (S_t - S_{t-1})/S_{t-1} = S_t/S_{t-1} - 1$ where S_t (S_{t-1}) is the current (previous) price of the asset and the relationship $r^c = \ln(1 + r^d)$ holds. The rational behind this approach to assume the prices to be log-normally distributed: A random variable S is said to be log-normally distributed if $S = a + c \cdot e^z$ where a and c are constants and z is normally distributed. Alternatively, a random variable J is said to be S_U–normal if $\sinh^{-1}(J) = \lambda + \theta z$ where λ and θ are constants and z, again, is normally distributed. This latter distribution, suggested by Johnson (1949), is a very flexible skewed and leptokurtic distribution and is therefore also well apt for modeling asset returns.

A different way of modeling higher moments would be the introduction of dynamics. By splitting the considered period into several (or infinitely many) subperiods, a discrete (continuous) price process can be modeled that reflects a *random walk*. The arguably most popular of these processes is based on the *geometric Brownian motion* for a price path of $S_{t+\Delta t} = S_t \cdot \exp(\alpha \cdot \Delta t + \sigma \cdot dz)$ where $E(S_{t+\Delta t}) = S_t \cdot \exp\left(\alpha \cdot \Delta t + (\sigma^2/2) \cdot \sqrt{\Delta t}\right)$ and the *Wiener process* $dz = \lim_{\Delta t \to 0} \sqrt{\Delta t} \cdot \varepsilon_{t+\Delta t}$ with $\varepsilon_{t+\Delta t} \sim N(0,1)$. Higher moments can then be depicted when α and, in particular, σ^2 are no longer kept constant but are assumed to change over time or to be stochastic,

---

[18] See also Mandelbrot (1997) who stimulated the debate whether the second moment, the variance, is finite for all time series.

[19] See, e.g., Tsiang (1972), Francis (1975), Friend and Westerfield (1980) or Scott and Horvath (1980).

[20] For more recent results, see Dittmar (2002) and the literature quoted therein.

too. The resulting models are demanding[21] and are hence applied merely in fields such as *Financial Engineering* or *EconoPhysics*, e.g., for pricing single assets, yet are rarely applied when a larger group of assets is to be investigated at the same time.

If these technical difficulties are to be avoided, a less demanding (and quite popular) approach is to assume the validity of the central limit theorem according to which the combination of random variables will have a probability that is close to the normal if the number of such variables is sufficiently large and if the variables themselves are independent identically distributed (*iid*) and have non-zero finite variance. This assumption is often made for portfolios where the returns are sufficiently close to the normal even when the returns of the included asset are not. However, more recent results indicate that portfolios' returns are not closer to normality than the included assets' – in particular when the assumption of iid returns is violated and, e.g., a portfolio contains both derivates and the underlying assets. As will be shown in chapter 6, asset selection with a mere focus on expected (utility of) return and risk can provoke and enforce undesired properties of the resulting portfolio's return distribution, in particular when and inappropriate risk measures are applied. Likewise, abandoning the normality assumption and introducing empirical distributions might turn out as a Pandora's Box: the parametric distribution might be superior even when it is known that it is not stable, as will also be shown in chapter 6.

In the recent literature, dynamic aspects have gained considerable interest. Not least because of the theoretical complexity, the respective models have to rely on strong assumptions in order to remain manageable: Discrete time models are often tackled with *dynamic programming* approaches[22] starting with solving the myopic problem for the last period and then, step by step, solving the problem for the preceding period.[23] However, not all dynamic optimization problems are manageable with this approach: not all dynamic problems can be "chopped" into a series of myopic problems without destroying relevant dynamic aspects; the number of sub-

---

[21] For a presentation of these methods, see Brooks (2002) or Campbell, Lo, and MacKinlay (1997).

[22] See Hillier and Lieberman (1995) and section 2.1.1.

[23] See, e.g., Huang and Litzenberger (1988, section 7.9).

problems might exceed the computational resources, and not all continuous-time problems can be transferred into a discrete-time framework.[24]

Extensions to the traditional MPT models by relaxing assumptions on frictionless markets, including the introduction of transactions costs, allowing for wholenumbered quantities rather than infinitely divisible stocks, upper limits on the number of different stocks, non-parametric distributions, alternative risk measures etc., cause severe computational problems, some of which will be discussed in the following chapters. The traditional finance literature therefore has either to exclude these aspects or be rather selective in incorporating frictions, frequently at the cost of excluding other aspects relevant in practice. Chapters 3 through 6 will discuss some of these extensions, present some attempts from the literature and offer new answers by original work.

1.1.3   Risk Reconsidered

1.1.3.1   Definitions of Risk

So far, the term *risk* has been used to characterize a situation where the exact outcome is not known and where the employed risk measure indicates the magnitude of deviations from the expected value. "Risk" therefore reflects not only the "dangers" associated with an investment, but also the chances; in this sense, a risky situation is one in which surprises and unexpected developments might occur. A typical representative of this type of risk measures is the volatility of returns which not only is one of the foundations of portfolio theory as introduced so far, but will also be the prime measure of risk for most of the remainder of this contribution.

Alternatively, risk can denote the fact that things "can go wrong" and losses or underachievement can be incurred. For this notion of risk, *semi-variance* is an early example which measures only the negative deviations from the expected value. More recently, *Value at Risk* has become a very prominent figure. It denotes the maximum

---

[24] See Merton (1992) or Duffie (2001). Cvitanić and Karatzas (1999) discuss dynamic risk measures. Sundaresan (2000) gives a concise review of the major models. For literature on continuous portfolio selection, see, e.g., Loewenstein (2000), Nilsson and Graflund (2001), Brennan and Xia (2002), Andersen, Benzoni, and Lund (2002) or Gaivoronski and Stella (2003) and the literature quoted therein. An alternative view is offered by Peters (1996) who applies chaos theory.

loss within a given period of time with a given probability. Chapter 6 will be concerned with the concept (and, notably, the shortcomings) of this risk measure. A third notion of risk refers to a situation of danger or peril. In finance, this concept is applied, e.g., in Stress Testing where the effects in extreme and undesired situations are investigated. This aspect of risk as well as downside risk measures are merely applied in circumstances where catastrophes ought to be prevented, investors ought to be saved from hazardous situations, or additional or supplementary information on an investment is wanted. This includes tasks of supervising bodies and regulatory authorities as well as issues in corporate and financial governance. Much like measures that focus only on the positive surprises and upside risk, the associated risk measures, however, appear less appropriate as a prime (or sole) risk measure in financial management.

Risk can also be characterized by its sources, fundamental causes and effects. This contribution is mainly concerned with

financial risk where the future value of an asset or portfolio is not known exactly and the potential outcomes are associated with uncertainties.

This might include or add to other types of risks, such as

inflation or price-level risks which reflect that the future purchasing power of a certain amount of money is unknown;

exchange risk in cases where investments include assets in foreign currencies and the future exchange rates might be uncertain;

interest rate risks since the prices of many assets are linked to the general interest rate, (unanticipated) changes in which will therefore move these assets' prices into (unanticipated) directions;

political and legal risks which affect the assets' issuer and his ability to carry on his businesses as planned and might have a severe effect on the issuers operations or profit situations when the legal situations or the tax systems are changed;

credit risk, default risk of major suppliers and customers which can affect the profit, production, or distribution;

to name just a few. Hence, there also exist risk measures that asses the current exposure to an undesired situation and the disadvantages resulting from it. A typical representative for this type would be a bond's *duration* that measures the price changes due to changes in the interest rate (i.e., the bond's interest rate risk).

Arguably, volatility is the most widespread risk measure in financial management – and is often even used synonymously for the term risk.[25] The reasons for this are various, most notably among them perhaps the fact that it reflects the assumption of normally distributed returns or of price processes with normally distributed price movements. Also, it measures not just potential losses but the whole range of uncertain outcomes and therefore serves the demands on a "general purpose" risk measure in finance. Not least, it also has a number of favorable (technical and conceptual) properties. Hence, this contribution, too, will use volatility as preferred measure of risk.

1.1.3.2   Estimating the Volatility

The concept of volatility is rather simple and straightforward: Since the variance measures the average squared deviation of the expected value, the volatility, calculated by taking the square root from the variance, allows for ranges or bandwidths around the expected value within which the actual value will be realized with a certain probability. Estimating the actual value for an asset's volatility, however, is not always as simple and straightforward. A more fundamental problem with the implicit approach is that it demands market participants to have knowledge of the true future volatility – which, of course, might be a rather daring assumption. Or, in the words of Lewellen and Shanken (2002):

> "Financial economists generally assume that, unlike themselves, investors know the means, variances, and covariances of the cash-flow process. Practitioners do not have this luxury. To apply the elegant framework of modern portfolio theory, they must estimate the process using whatever information is available. However, as [Black (1986)] so memorably observes, the world is a noisy place; our observations are necessarily imprecise."

---

[25] For a critical discussion of commonly used risk measures in finance, see, e.g., Szegö (2002) and Daníelsson (2002).

There are several approaches to estimate the volatility, all of which have their advantages and disadvantages. In theory as well as in practice, two major groups of approaches have gained considerable popularity:

Implicit approaches search for values of the volatility that justify currently observed prices for securities the price of which depends on the volatility and under the assumption of valid (and actually employed) equilibrium pricing models.

Historic approaches are based on past time series and assume that history will repeat or will be continued. The past ("in sample") data are thought well apt to characterize future ("out of sample") developments. Hence, it is assumed that the historic, in sample volatility allows a reasonable estimate for the out of sample volatility.

A typical way of estimating the implied volatility of a stock would be to find the value that "explains" the current price of a buy (*call*) or sell (*put*) option on this stock. These derivative securities depend on their time to maturity, their exercise price (at which the underlying can be bought or sold), the safe interest rate, the current stock price, possible dividend payments and dates before maturity – and the underlying stock's volatility which is the only variable that is neither specified in the option contract nor directly observable from market data. With exact option pricing formulas or analytic approximation models,[26] the *implicit volatility* can be found by searching the value for σ where the empirically observed (i.e., quoted) price equals the theoretical price. Classical numerical search methods are usually able to find the implicit value for σ.

The implicit approach is sometimes considered superior as it does not require historic data and is able to capture current market expectations that are not (or cannot be) represented by past observations. However, this approach has also some severe shortcomings. Apart from "plain vanilla" type options, option pricing models tend to be rather demanding, and their implementation might easily turn into a real challenge reaching (or even exceeding) the capacities of standard spreadsheet applications and programming environments. The assumption that the quoted option price results from the (majority of the) market participants using these models

---

[26] For a comprehensive introduction to options and derivatives and pricing models, see Hull (2003).

might therefore not always be justified. This might contribute to phenomena such as the often observed *volatility smile*: other things equal, the more the underlying's current price and the option's exercise price differ, the higher the implicit volatility. In addition, most models assume constant volatility until maturity which is often seen as strong simplification, in particular for options that can be exercised prematurely. Not least, for the vast majority of financial assets, there exist no derivatives from which to estimate the implicit volatility. Hence, while suitable (and reasonably reliable) for assets on which derivatives exists and where both assets and derivatives are traded at a high frequency, the implicit approach is not generally applicable.

Historic approaches, on the other hand, are easier to apply when standard approaches are assumed, yet might become demanding when usual assumptions (such as stable volatility over time) are dropped and might become imprecise when there are not enough historic data, when there are structural breaks within the time series or when there is a fundamental event that affects the issuer's future situation but is not yet reflected in the (past) data used for estimating the volatility. Nonetheless, application of historic approaches is common practice in both theory and practical application, for one because they are the only alternative, but also because they are rather simple and have been found to work sufficiently well.

1.1.3.3 Estimation of the Historic Volatility

Given a series of past observations of an asset's returns, r_t, the straightforward way of estimating its volatility would be to calculate the in sample variance by

$$\hat{\sigma}^2 = \frac{1}{\tau - 1} \sum_{t=1}^{\tau} (r_t - \bar{r})^2$$

or, under maximum likelihood, by

$$\hat{\sigma}^2 = \frac{1}{\tau} \sum_{t=1}^{\tau} (r_t - \bar{r})^2$$

where $\bar{r}$ is the in sample average return and τ is the number of observations. For many instances, this approach is sufficiently exact; yet it might get imprecise when data from the too distant past ought to predict the near future[27] and when the general (market) situation has changed. Financial time series often exhibit what is called

---

[27] To take this into account, the data might be weighted where the more recent observations might get a higher weight than ancient ones.

volatility clustering, i.e., volatility is changing over time, and days with high risk are likely to be followed by days with high risk, and days with low risk are likely to be followed with days with low risk.

To account for this fact, models have been developed that abandon the assumption of constant volatility and allow for autocorrelation, i.e., correlation of a variable with its own past observations. *Generalized Autoregressive Conditional Heteroskedasticity (GARCH)* models[28] therefore assume that the current volatility depends not only on a constant term, but also on the p recent values for the volatility and on the magnitude of the q recent error terms in the data series which is also considered to exhibit autocorrelation. The specification for a GARCH(p,q) model therefore reads

$$\sigma_t^2 = \alpha_0 + \sum_{i=1}^{q} \alpha_i \cdot e_{t-i}^2 + \sum_{j=1}^{p} \beta_j \cdot \sigma_{t-j}^2$$

with

$$r_t = x_t'\gamma + e_t, e_t \sim N\left(0, \sigma_t^2\right)$$

where x_t and γ are the explanatory variables and regression parameters, respectively, for the returns.[29]

Under the assumption of conditionally normally distributed errors, there is no closed-form solution for the parameters $\alpha_0, \alpha_i, \beta_j$ and γ. They can be determined, though, by maximizing the respective log-likelihood function which, for a GARCH(1,1), is

$$\mathcal{L} = -\frac{T}{2}\ln(2\pi) - \frac{1}{2}\sum_{t=1}^{T}\ln\left(\sigma_t^2\right) - \frac{1}{2}\sum_{t=1}^{T}\frac{(y_t - x_t'\gamma)^2}{\sigma_t^2}. \tag{1.14}$$

Many statistics and econometrics software packages offer solutions to this problem by employing iterative methods. However, as the function (1.14) can have many different local maxima, the numerical methods employed by these software packages

---

[28] GARCH models where introduced independently by Bollerslev (1986) and Taylor (1986). For a more detailed presentation and preceding models such as ARCH (Autoregressive Conditional Heteroskedasticity) and ARMA (Autoregressive Moving Average) models, see, e.g., Brooks (2002, chapter 8) or Campbell, Lo, and MacKinlay (1997, chapter 12).

[29] Note that x_t might also contain lagged values of the return.

might produce different results when they start off with different initial parameter guesses.[30] In section 2.4, it will be demonstrated how the problem of parameter estimation for GARCH models can also be approached with heuristic optimization techniques.

1.1.3.4   A Simplified Approach: The Market Model

Frequently, estimating the volatilities needed for mean-variance based portfolio optimization is not as much of a problem of precision but of the rather large number of parameters involved in the model: When there are N different assets, then N variances and $(N^2 - N)/2$ different covariances need to be estimated. For a large number of assets, the estimation problems arising from the large number of parameters can quickly get out of hand. Sharpe (1963) suggests a simple, yet path breaking simplification by assuming that "the returns of various securities are related only through common relationships with some basic underlying factor"[31] and that the return of any security i can be described by the linear relationship with this factor's return

$$r_{it} = a_i + b_i \cdot r_{Mt} + e_{it} \tag{1.15}$$

where M is the factor of choice and e_{it} is the error term. Finding the variance of r_{it} becomes then equivalent to finding the variance of the right-hand side of equation (1.15): The intercept a_i is a constant and will therefore not contribute to σ_i^2. Also, if r_i and r_M are stationary and bivariate normal, then it follows that the error terms are uncorrelated with r_M, hence $\text{Var}(b_i \cdot r_M + e_i) = \text{Var}(b_i \cdot r_M) + \text{Var}(e_i)$. Furthermore, the error terms of different assets can be considered to be uncorrelated, $\text{Covar}(e_i, e_j) = 0$.

Based on these definitions and considerations and after some straightforward rearrangements, the simple, yet crucial result is that an asset's variance and its covariance with other assets can be estimated via the market's risk according to

$$\sigma_i^2 = b_i^2 \cdot \sigma_M^2 + \sigma_{e_i}^2,$$
$$\sigma_{ij} = b_i \cdot b_j \cdot \sigma_M^2.$$

---

[30]   See Brooks, Burke, and Persand (2001).

[31]   Cf. Sharpe (1963, section iv).

From a practical point of view, the Market Model makes the preliminary estimations in portfolio optimization less cumbersome: to find the different variances and covariances, one simply has to estimate σ_M^2 and for each of the N assets the parameters a_i, b_i, and $\sigma_{e_i}^2$ – which is considerably below the original $N + {(N^2 - N)}/{2}$ estimates necessary for the covariance matrix[32] and which can also be used as a simple means to generate artifical market data.[33]

Yet, there is another interesting result from this approach. The variance of i is split into two parts: $\sigma_{e_i}^2$ is called *non-systematic, unsystematic* or *diversifiable risk* because it can be avoided by holding an appropriate portfolio rather than the single asset. $b_i^2 \cdot \sigma_M^2$ is also known as *systematic risk* and denotes the part of the asset's volatility that cannot be diversified. With b_i coming from an Ordinary Least Squares (OLS) estimation, it can be determined by $b_i = \sigma_{iM}/\sigma_M^2$; after substituting the covariance's definition, $\sigma_{iM} = \sigma_i \cdot \sigma_M \cdot \rho_{iM}$, and simple rearrangements, the asset's variance can also be partitioned into

$$\sigma_i^2 = \underbrace{\rho_{iM}^2 \cdot \sigma_i^2}_{=b_i^2 \cdot \sigma_M^2} + \underbrace{\left(1 - \rho_{iM}^2\right) \cdot \sigma_i^2}_{=\sigma_{e_i}^2}. \tag{1.16}$$

This refines the previous finding from section 1.1.1: A portfolio with many different assets will have a variance closer to the average covariance than the average variance (akin to Figure 1.2), hence investors with large and well diversified portfolios should be merely concerned with the assets' covariances (or correlations) with the market and less with the assets' overall risk, i.e., variance, since not all of the variance will affect them.

1.2   Implications of the MPT and Beyond

1.2.1   The Capital Market Line

In the Tobin framework, the expected returns are exogenously given and the N simultaneous equations where solved for the weights. Under the assumption of homogenous expectations, all investors will hold the same optimal tangency portfolio

---

[32] For another simplified approach, see Markowitz and Perold (1981).

[33] Issues related with the modeling of variance and skewness in financial data series are presented in Liechty (2003).

with equal weights x_i: the different utility curves affect only the fractions for the safe asset and the tangency portfolio but not the composition of the tangency portfolio itself. This implies that a certain asset j is represented either in any investor's portfolio with the same proportion (within T) or in none; or in other words, all existing assets are held by all investors, since an asset an investor doesn't hold will not be included in any other investor's portfolio.[34] This then implies that in an equilibrium market the weights x_i must correspond to the fraction asset i represents of the market, $x_i^{\mathcal{M}}$. In equilibrium, the ideal tangency portfolio must therefore equal the market portfolio, i.e., $T = \mathcal{M}$ and $x_i = x_i^{\mathcal{M}}$.

Hence, the central relationship from the Tobin Model (see (1.11) on page 12) between a portfolio's risk and its return can be written as

$$r_P = r_s + (r_{\mathcal{M}} - r_s) \cdot \frac{\sigma_P}{\sigma_{\mathcal{M}}}. \tag{1.17}$$

The linear relationship between the expected return of a portfolio and its volatility expressed in Equation (1.17) is generally referred to as *Capital Market Line* (*CML*).

An important point is that for the derivation of the CML, short-sales are not excluded as models with limits on the weights and risk-free borrowing or lending, such as in Markowitz (1952) or Dyl (1975), will not lead to equilibrium asset pricing models; cf. Alexander and Francis (1986, p. 125). However, Lintner (1971) showed that allowing short-sales is a mathematically convenient, yet not a necessary assumption. The intuitive argument is basically that no asset can have negative weight in the equilibrium market portfolio since every investor would want to sell the asset but none would want to buy it; if the market portfolio actually contained assets with negative weights, then markets wouldn't clear.

The underlying concept has also become a standard measure for *ex post* portfolio evaluation, based on Sharpe (1966). Rearranging (1.17) yields

$$\frac{r_P - r_s}{\sigma_P} = \frac{r_{\mathcal{M}} - r_s}{\sigma_{\mathcal{M}}}. \tag{1.18}$$

---

[34] Empirical studies confirm this view and show that funds with *passive investment strategies* (which try to replicate some market index) are superior to most *actively managed* funds (i.e., where managers have expectations diverging from the other market participants' and therefore prefer other weights); see Sharpe, Alexander, and Bailey (2003).

When r_M and σ_M are realized and observable *ex post*, then the slope of the *ex post* CML (1.18) is the achieved *reward to variability ratio* or *Sharpe Ratio (SR)*[35] of the portfolios M and P, respectively. The *SR* corresponds directly to the slope of Tobin's efficient line in equation (1.12) on page 12, and with the same argument, investors will be better off the higher the Sharpe Ratio of their investments:

$$SR_{P^*} > SR_P \Rightarrow P^* \succ P. \tag{1.19}$$

Due to its popularity as an *ex post* criterion for portfolio evaluation, *Sharpe Ratio* has also become a common synonym for the (*ex ante*) slope of the Tobin efficient line.

1.2.2 Capital Asset Pricing Model

As argued previously, rational investors with homogeneous expectations will all hold some combination of the market portfolio, M, and a safe asset with a return of r_s. Given the proportions x_i^M, the variance of the market portfolio is

$$\sigma_M^2 = \sum_i \sum_j x_i^M \cdot x_j^M \sigma_{ij}.$$

Making use of the statistical property $\sum_j x_j^M \sigma_{ij} = \sigma_{iM}$, it can also be written as

$$\sigma_M^2 = \sum_i x_i^M \cdot \sigma_{iM} = \sum_i x_i^M \cdot \sigma_i \cdot \sigma_M \cdot \rho_{iM}$$

$$\Rightarrow \sigma_M = \sum_i x_i^M \sigma_i \cdot \rho_{iM}.$$

The variance of the market portfolio is therefore the weighted sum of the assets' covariances with M. Hence, it is not necessarily high asset volatility that contributes to the portfolio's risk, but large covariance between assets and the market as this is the part of the volatility that cannot be diversified, $\sigma_i \cdot \rho_{iM}$. The investor therefore expects assets with higher undiversifiable or *systematic risk* to contribute more to the portfolio's expected return than assets with low systematic risk.[36]

---

[35] See also Sharpe (1994).

[36] See also equation (1.16) in section 1.1.3.4.

Given these considerations, the market will be in equilibrium only if the ratio between contributed return and contributed systematic risk is the same for all assets. If there was an asset that would pay a lower return for same additional risk, it would not be held in any portfolio since its inclusion would lower the portfolio's return/risk ratio; assets with over-proportionate returns would be highly demanded as they would increase the portfolio's return/risk ratio. In either case, the market would therefore be out of equilibrium: demand and supply would not balance, and the market would not clear. With similar arguments, assets that have no systematic risk should earn exactly the safe rate of return. Using the market's expected return as a benchmark and by $\sigma_{MM} = \sigma_M^2$, the equilibrium expected return for an asset i is therefore

$$\frac{r_i - r_s}{\sigma_{iM}} = \frac{r_M - r_s}{\sigma_{MM}}$$

or, in the usual notation

$$r_i = r_s + (r_M - r_s) \cdot \beta_i \tag{1.20}$$

with

$$\beta_i = \frac{\sigma_{iM}}{\sigma_M^2} = \frac{\sigma_i \cdot \rho_{iM} \cdot \sigma_M}{\sigma_M \cdot \sigma_M} = \frac{\sigma_i \cdot \rho_{iM}}{\sigma_M}. \tag{1.21}$$

The *Capital Asset Pricing Model (CAPM)*, independently developed by Sharpe (1964), Lintner (1965) and Mossin (1966), therefore states that in equilibrium there is a linear relationship between the systematic risk and the risk premium. Using the beta coefficient as the relevant risk measure as defined in (1.21), all asset should form a line in the return-beta space according to the *Security Market Line (SML)* in equation (1.20).[37] When considering expected values for a market in equilibrium, only the systematic risk should affect the risk premium. The unsystematic risk can be avoided (hence also referred to as diversifiable risk), an investor bearing it therefore doesn't "deserve" an additional risk premium.

The validity of the CAPM was to be tested and confirmed in a number of empirical tests for many markets (most notably by Black, Jensen, and Scholes (1972), Fama and MacBeth (1973), and Blume and Friend (1975)) which were eventually heavily

---

[37] A more rigorous presentation of the underlying mathematics can be found in, e.g., Huang and Litzenberger (1988); for a presentation of the theory of asset pricing, see, e.g., Duffie (2001).

criticized by Roll (1977). Roll's critique is that the use of a market proxy for the true (yet unknown) market portfolio prohibits empirical tests; Roll, however, does not reject the CAPM per se nor its testability in theory.[38] Consecutive studies develop alternative methods but are still undecided whether the CAPM is valid or not: Levy (1978) acknowledges the fact that the average investor will not hold the market portfolio but a very small number of different assets for various reasons and finds that for widely held stocks, the beta is an appropriate risk measure whereas for others the variance is more relevant. Cheng and Grauer (1980) find evidence against it yet concede that their results might also be interpreted as evidence against the stationarity assumption of the return distributions; Gibbons (1982) rejects two versions of the CAPM[39] for the New York Stock Exchange whereas Stambaugh (1982) finds evidence in favor for the zero-beta CAPM. Other authors show the importance of alternative aspects such as firm size[40] or survivor bias[41]. Levy (1997), in response, emphasizes that empirical *ex post* results are not enough to ultimately reject the (over time instable) *ex ante* parameters of the CAPM and concludes from his experiments "that the risk-return equilibrium model is not dead; it is alive and doing better than previous empirical studies have revealed."

The question of whether the CAPM is valid or not is therefore still far from being answered:[42] Whereas some see "damning evidence" (as does Haugen (2001, p. 249)), others concede that "although we couldn't tell whether it was true or not, it does give us insight into behavior in capital markets" (cf. Elton, Gruber, Brown, and Goetzmann (2003, p. 358)).

---

[38] See also Roll (1978) as well as Lo and MacKinlay (1990) and Lewellen and Shanken (2002).

[39] In addition to the standard CAPM as presented, the *zero-beta CAPM* is tested: If both the CAPM and the MM hold, an asset or portfolio with $\beta = 0$ ought to earn r_s. This (efficient) zero-beta portfolio then replaces the (missing) safe asset, making use of the property that any combination of two efficient portfolios will again be efficient (provided short sales are allowed when necessary).

[40] See Fama and French (1992).

[41] See the opposing results from Kothari, Shanken, and Sloan (1995) and Fama and French (1996).

[42] For a more in-depth discussion of the literature, see, e.g., Alexander and Francis (1986) or Sharpe, Alexander, and Bailey (2003).

1.2.3 Alternative Versions of the CAPM

Because of the problems associated with its empirical application, alternatives to the standard CAPM have been presented which contain either extensions or modifications to the original version, or are based on alternative paradigms. Merton[43] relaxes the myopic nature of the CAPM and introduces a continuous time version, *Intertemporal CAPM (ICAPM)*, where there is a safe return[44] and the prices of the risky securities follow a diffusion process. In particular, he assumes that investors are concerned with market risks as well as "extra-market sources of risks" which are referred to as *factors*. He consequently suggests a *Multifactor CAPM* where the asset's risk premium depends not only on the market's risk premium but also on these factors' risk premia:

$$r_i = r_s + (r_M - r_s) \cdot \beta_{iM} + \sum_f (r_f - r_s) \cdot \beta_{if}$$

where r_f is the expected return of factor f and β_{if} is the sensitivity of asset i to this factor. As different investors might be concerned with different extra-market factors and might therefore have different strategies to hedge their respective relevant risks, translating this model into practice or testing is not straightforward. Recently, Campbell and Vuolteenaho (2003) operationalize the ICAPM by suggesting a two beta CAPM: They break the market beta into two components which reflect the market's expectations about future cash flows ("bad" cash flow beta) and about future discount rates ("good" discount-rate beta), respectively. Based on this distinction, the authors find that small stocks have higher cash flow betas (and hence higher returns) than large stocks and growth stocks; this might not only explain the *size effect*, a common market anomaly, but also helps explain why the standard CAPM is found to perform rather poorly for the last decades.

Breeden (1979) presents a Consumption-Based Continuous-Time CAPM where the beta-coefficient is measured relative to aggregate consumption, and the change in aggregate consumption is used rather than the market return. If marginal utility of consumption will be high (low) in good (bad) times, assets with payoffs that are positively correlated with aggregate consumption will not be as attractive as those

---

[43] See Merton (1969, 1971, 1972, 1973).

[44] This assumption is relaxed in Richard (1979).

which are negatively correlated as the latter give their highest payoffs when they are "most needed," namely in "bad times." Consequently, assets with positive (negative) correlation will have relatively low (high) present values – which results in high (low) expected returns. Friend, Landskroner, and Losq (1976) also take inflation into account and derive a multi beta CAPM where the additional price of inflation risk enters the equilibrium nominal price.[45] How intertemporal asset pricing models can be evaluated econometrically, is discussed in Hansen, Heaton, and Luttmer (1995).

Grossman and Stiglitz (1980) present a model that takes into account information gathering costs, and Admati (1985) shows that in this case decisions will still be made within the mean-variance framework, yet each investor will have an individual "market portfolio" which will be her individual benchmark and therefore contribute to individual parameters for the CAPM.

An aspect frequently neglected in pricing models (as in the original CAPM) is the effect of taxes. A first major breakthrough was presented by Brennan (1970) who investigates the case where investors have different tax rates. He finds a modified version of the CAPM where the expected return depends on the beta on the price changes as well as on the dividend; hence the original CAPM's security market line is replaced with a security market plane. When dividends (together with safe returns) are taxed at a different tax rate than capital gains, then a "dividend clientele effect" can be observed: As investors will be concerned with their after-tax returns, companies are likely to be owned by stockholders who are attracted by the respective dividend policy with regard to the investors' tax classes.[46]

Another concern is that in the long run, there will be hardly an asset that is perfectly risk-free. One suggested solution to this problem is the zero-beta CAPM, where an asset or portfolio with no systematic risk, i.e., with $\beta = 0$, is used for an substitute: As can readily be seen from equation (1.20) of the SML, this zero-beta asset (or portfolio, respectively) should earn no risk premium, and its yield should reflect the supposed level of return for safe investments under the given market situation.[47] Gilster, Jr. (1983) shows that in the absence of a risk-free asset, the com-

---

[45] See also Elton and Gruber (1984), Halmström and Tirole (2001), and Brennan and Xia (2002).

[46] Brennan's results, however, stirred some discussion and counterarguments; a short review of the main literature can be found, e.g., in Alexander and Francis (1986, section 8.3A).

[47] See also footnote 39 on page 28 and the more detailed presentation in, e.g., Elton, Gruber, Brown, and Goetzmann (2003, pp. 310-320).

position of the efficient frontier is the same for any investment horizon. Also, the existence of non-marketable assets implies that an investor might hold a combination of publicly available securities as well as individual investments. In particular those stocks from the market portfolio that have a high correlation to these non-marketable assets will therefore have a lower weight in the investor's portfolio than in the market portfolio.[48] Gonedes (1976) finds the beta is a sufficient measure of risk even when investors have heterogeneous expectations.[49] Lindenberg (1979) derives equilibrium conditions when investors are price affecters rather than price takers only.

Since empirical investigations show that betas tend to be unstable over time and might even be random coefficients when estimated with an OLS regression[50], there are also alternative versions for estimating the beta coefficient itself. Rosenberg and Guy (1976) were the first to distinguish industry betas and allow for adjustments of the short-term beta forecasts based on investment fundamentals and balance sheet information.[51] Grossman and Sharpe (1984), too, estimate an asset's beta via a factor model which consists of a fixed term depending on the asset's industry and the weighted size, yield, and past beta.[52] Gençay, Selçuk, and Whitcher (2003) suggest the use of wavelet variances for more precise beta estimates.

All of these extensions to the CAPM have improved the reliability of estimations and have helped to understand important aspects of capital markets. However, despite the shortcomings of the original CAPM when applied empirically, it is a common means in computational studies for generating data sets based on empirical observations. While historic volatilities and covariances can be used for estimates of the future risk,[53] historic returns cannot simply be extrapolated into the future as the last period's return is not necessarily a good estimator for its expected value in the next period. For most of the computational studies in this contribution, the original CAPM will be used for generating plausible data for a market in equilibrium; for

---

[48] See Mayers (1972), but also Mayers (1973) where non-marketable assets together with the absence of the risk-free asset are investigated.

[49] See also Grossman and Stiglitz (1976).

[50] See, e.g., Fabozzi and Francis (1978) and Shanken (1992).

[51] See also Rosenberg (1985).

[52] See also section 16.B in Sharpe, Alexander, and Bailey (2003).

[53] See also sections 1.1.3 and 1.2.1.

this purpose, extended or alternative models would not lead to additional insights as will be seen, e.g., in chapter 4. The CAPM as presented in equations (1.20) and (1.21) will therefore be used in conjunction with empirical data for major capital markets.

1.2.4   The Arbitrage Pricing Theory

Whereas mean-variance analysis and certain assumptions on the investors' behavior are the foundation for the Markowitz and subsequent models, alternative approaches are based on other assumptions and relationships. One of these is the *law of one price*: Any two investments with equal future claims and payoffs must have the same price today. Assuming a perfect market where the number of possible future states equals the number of (linearly independent) securities, than the future payoff structure of any new asset can be replicated with an adequate portfolio of existing (and already priced) securities. If the price for the new asset would differ from the price for the replicating portfolio, arbitrage would be possible: the investor could buy the asset while going short in the portfolio (or *vice versa*) and all future payments would perfectly offset each other. Since all future net payments of this combination will be zero regardless of the actual market situation, then the net price of this combination must not be positive.

The *Arbitrage Pricing Theory (APT)* by Ross (1976) is built on this consideration. The central assumption of the APT is that asset i's return, r_i, has a linear relationship to the returns of the factors $f \in \mathcal{F}, r_f$, of the form

$$r_i = b_{i0} + \sum_{f \in \mathcal{F}} b_{if} \cdot r_f + \varepsilon_i. \tag{1.22}$$

Ideally, the error term ε_i has zero variance, yet it is sufficient to assume that the number of factors is large and the residual risk is small enough.[54] Also, it is assumed that it is uncorrelated with any other asset's error term as well as the factor returns. The factors themselves can be regarded as portfolios from the existing assets. Ideally, these factors are uncorrelated to each other which facilitates the interpretation of the b_{if}'s as sensitivities of r_i towards the factors.[55]

---

[54]   See, e.g., Dybvig (1983), Grinblatt and Titman (1983, 1985) and Ingersoll, Jr. (1984). MacKinlay and Pástor (2000) discuss the effects of missing factors.

[55]   A set of correlated indices can be transformed into a set of uncorrelated indices; see Elton, Gruber, Brown, and Goetzmann (2003, section 8.A). Hence, this assumption is useful, yet not necessary for the validity of the APT.

Taking the expected value of equation (1.22) and applying some algebra,[56] the APT can be written as

$$E(r_i) - r_s = \sum_{f \in \mathcal{F}} b_{if} \cdot \left(E(r_f) - r_s \right). \tag{1.23}$$

where r_s is the risk-free rate of return. Hence, according to the APT, the expected return of an asset ought to consist of the safe return plus a risk premium depending on the sensitivity towards the expected "risk premium" of some factors. Here, it also becomes apparent that the CAPM can be viewed as a special case of the APT where the only factor is the market portfolio.[57]

The APT has several advantages over the CAPM: Though both equilibrium models assume perfect markets and concave utility functions, the APT does not require the additional assumptions made in the CAPM; in particular, it does not demand the existence and knowledge of the (unobservable) market portfolio. A number of empirical tests[58] found that the APT is superior to the CAPM in terms of explanatory power and reliability of predictions. At the same time, there is no conclusive general result on the selection of the factors. Chapter 7 of this contribution will present new results for the factor selection problem; a more detailed presentation on the relevant literature on the APT is therefore left to section 7.1.

1.3 Limitations of the MPT

Modern Portfolio Theory has become not only the basis for theoretical work on portfolio optimization but also a major guideline for institutional portfolio management. Direct application, however, of the theoretical results to practical problems is not always possible for various reasons. First and foremost, the underlying assumptions and constraints are mostly chosen in a way to make the models solvable yet often at the cost of strong simplifications of real market situations. Without these simplifications and stylized facts, however, the capacities of traditional methods are

---

[56] For a more rigorous presentation, see, e.g., Lehmann and Modest (1987).

[57] See Sharpe, Alexander, and Bailey (2003, section 11.5).

[58] See in particular Roll and Ross (1980). Shukla (1997) surveys relevant literature.

quickly exceeded. With new methods at hand, these restrictions can be overcome, as will be shown in the main part of this contribution.

There are a considerable number of problems which resist parameterization or where a formal model would be bound to cancel out relevant aspects. Also the there has been raised serious critique on quantitative methods and their implicit assumptions and consequences.[59] In this respect, new approaches including *Economic Realism*[60] or *Behavioral Finance*[61] make significant contributions to current research.

Some authors criticize the assumptions on the investors' behavior: Empirical studies as well as experiments show that investors' behavior is not necessarily rational but can be severely irrational: Kroll, Levy, and Rapoport (1988), e.g., found in their experiments that the participants failed to make investment decisions as predicted by the separation theorem; however, performance improved as the reward was increased tenfold. Weber and Camerer (1992) report from their experiments that investors tend to select suboptimal portfolios which differ from the (theoretically expected) market portfolio, trade too frequently, and might even select portfolios with negative expected returns.[62]

Another point of critique is the usual assumption that investors base their expectations about return merely on the associated risk. In practice, however, predictions (and eventually investment decisions) are often based on *technical analysis* where estimates for future prices and price movements are derived, e.g., from patterns in past stock prices and trading volumes, preferably from graphical representations (*chart analysis*).[63] Though in use for more than a century (first attempts can be traced back to Charles H. Dow in the 1890's), it has been widely ignored or rejected by academia, mainly on two reasons: (i) if relevant information could be gained by looking at past prices, everybody would do it and current prices should immediately

---

[59] See, e.g., McCloskey (1998, 1996).

[60] See Lawson (1997).

[61] See Thaler (1993), Goldberg and von Nitzsch (2001) or Shefrin (2001).

[62] See also Shiller (2000) and Oehler (1995).

[63] See, e.g., Kahn (1999) and Lo, Mamaysky, and Wang (2000), yet also Sullivan, Timmermann, and White (1999)

contain this information;[64] (ii) technical analysis does not look at the fundamental values of the underlying companies. Recent results, however, show that "the conclusions reached by many earlier studies that found technical analysis to be useless might have been premature"[65] and that even rather simple rules might generate buy and sell signals that can lead to higher returns which cannot be explained with the generally agreed econometric and financial methodology. Models for simulating asset price behavior as a result from investor behavior such as *agent based models* therefore allow for a larger variety of investment styles.[66]

Behavioral financial economists also criticize the standard assumptions on how investment decisions are made. In contrast to the often assumed utility maximizing individual with rational expectations, investors are not a homogenous group,[67] let alone is there a representative investor.[68] Some authors trace investors' behavior back to different trading styles and distinguish "smart money" (i.e., literate investors) and "noise traders" (i.e., trend followers)[69] Hence, herding behavior can be observed which might eventually lead to *bubbles* where demand drives the price far beyond the actual fundamental value. With usually less dramatic consequences for whole financial markets, individual often have expectations that deviate from those of other market participants. Frequently, this results in an *active* investment strategy where (both individual and professional) investors hold portfolios which do not reflect the market portfolio. Empirical studies, however, repeatedly find that on the long run these actively managed funds are outperformed by *passive* investment strategies where a market index is sought to be replicated. The disadvantage of unjustified and wrong expectations is often worsened by transactions costs and frequent trading: individual investors are therefore prone to "pay a tremendous performance penalty for active trading."[70]

---

[64] In terms of the concept by Fama (1970), the *weak information efficiency* should hold where all past information is contained in the prices. The *semi-strong efficiency* requires any publicly available information to be included in the prices, and under *strong efficiency*, also non-public information is reflected in the prices.

[65] Cf. Brock, Lakonishok, and LeBaron (1992, p. 1757).

[66] See LeBaron (1999).

[67] See, e.g., Campbell and Kyle (1993).

[68] See, e.g., Kirman (1992).

[69] See, e.g., Shiller (1989).

[70] Cf. Barber and Odean (2000, p. 773). See also section 1.2.1.

These aspects are related to difficulties that emerge when information has to be distinguished from rumors or noise[71] and when "attracting attention" is the main trigger for an investment: Barber and Odean (2003), e.g., find in an empirical study on some 700 000 individual investors and 43 professional money managers that individual investors tend to be net purchasers of stocks that attract high attention, either by high abnormal trading volume, previous extreme price movements, or by (not necessarily positive) reports on these stocks in the news – and therefore differ to some extent from institutional investors. Also, purchases on these "high-attention days" are likely to underperform. Alternatively, a misperception of risk and the probabilities at which extreme events occur might influence the decision process, as Kunreuther and Pauly (2004) find.

In practice, investors seem to prefer simplified investment rules. One reason for this might be rather straightforward: Theoretical models are often to demanding when applied to large problems. As argued already, estimating a market's complete covariance matrix might be extremely time consuming (if not even impossible), the optimization problems (in particular when market frictions exist) are not solvable with available software, and not all requests (such as a "low" portfolio turn-over rate during a given period of time) can be included satisfactorily in a formal optimization model. Hence, *Asset Class Management*, e.g., simplifies the selection process by splitting the universe of available assets into (sub-)groups of related or similar securities, selecting the best of each group by some rule of the thumb or applying the formal model only on these preselections.[72]

Though behavioral finance has attracted considerable research interests and made valuable contributions to the understanding of capital markets, there are no generally agreed evaluation methods, it is hard to test these models[73] and many important aspects have yet to be addressed systematically.[74]

---

[71] See in particular Black (1986), whose contribution on *noise* initiated a major stream of research.

[72] See Farrell, Jr. (1997) and Gratcheva and Falk (2003), but also Lettau and Uhlig (1999) and Cesari and Cremonini (2003).

[73] See Campbell (2000)

[74] See also Stracca (2004) and van der Saar (2004).

1.4 Summary

This chapter presented a short introduction to some of the foundations of Modern Portfolio Theory (MPT) in the tradition of Harry Markowitz, James Tobin and William Sharpe. With respect to computability, these models have to rely on rather strict assumptions that are not always able to depict real market situations. Subsequent models try to include these missing aspects, yet suffer from other shortcomings as they usually have to make strong simplifications in other aspects in order to remain solvable.

In the main part of this contribution, too, the original MPT models will be enhanced to allow a more realistic study of portfolio management problems. Unlike other approaches in the literature, however, the trade-off between model complexity and its exact solvability will not be answered by "exchanging" one simplifying constraints for another, but by applying new solution methods and optimization techniques. The basic concepts of these methods will be presented in the following chapter.

Chapter 2

Heuristic Optimization

2.1   Introduction

2.1.1   The Problems with Optimization Problems

Optimization problems are concerned with finding the values for one or several decision variables that meet the objective(s) the best without violating the constraint(s). The identification of an efficient portfolio in the Markowitz model (1.7) on page 7 is therefore a typical optimization problem: the values for the decision variables x_i have to be found under the constraints that (i) they must not exceed certain bounds ((1.7f): $0 \leq x_i \leq 1$ and (1.7e): $\sum_i x_i = 1$) and (ii) the portfolio return must have a given expected value (constraint (1.7c)); the objective is to find values for the assets' weights that minimize the risk which is computed in a predefined way. If there are several concurring objectives, usually a trade-off between them has to be defined: In the modified objective function (1.7a*) on page 9, the objectives of minimizing the risk while maximizing the return are considered simultaneously.

The Markowitz model is a well-defined optimization model as the relationship between weight structure and risk and return is perfectly computable for any valid set of (exogenously determined) parameters for the assets' expected returns and (co-)variances (as well as, when applicable, the trade-off factor between portfolio risk and return). Nonetheless, there exists no general solution for this optimization problem because of the non-negativity constraint on the asset weights. Hence, there is no closed form solution as there is for the Black model (which is equal to the

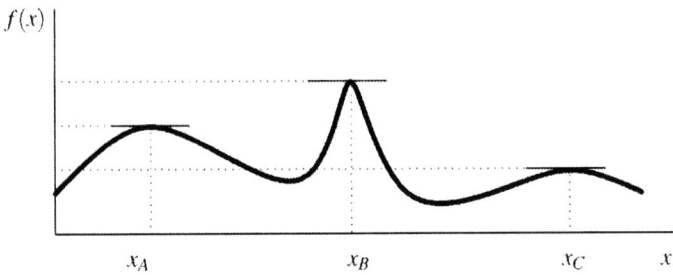

Fig. 2.1: Global and local optima

Markowitz model except for the non-negativity constraint). Though not solvable analytically, there exist numerical procedures by which the Markowitz model can be solved for a given set of parameters values.

Depending on the objective function, optimization problems might have multiple solutions some of which might be local optima. In Figure 2.1, e.g., a function $f(x)$ is depicted, and the objective might be to find the value for x where $f(x)$ reaches its highest value, i.e., $\max_x f(x)$. As can easily be seen, all three points x_A, x_B, and x_C are (local) maxima: the first order condition, $f'(x) = 0$, is satisfied (indicated by the horizontal tangency lines), and any slight increase or decrease of x would decrease the function's value: $f(x) \geq f(x \pm \varepsilon)|_{\varepsilon \to 0}$. Nonetheless, only x_B is a *global optimum* as it yields the highest overall value for the objective function, whereas x_A and x_C are just *local optima*. Unlike for this simple example, however, it is often difficult to determine whether an identified solution is a local or the global optimum as the solution space is too complex: All of the objective functions that will be considered in the main part of this contribution have more than one decision variable, the problem space is therefore multidimensional; and the objective functions are mostly discontinuous (i.e., the first derivatives are not well behaved or do not even exist).

In portfolio management, these difficulties with the objective functions are frequently observed when market frictions have to be considered. To find solutions anyway, common ways of dealing with them would be to either eliminate these frictions (leading to models that represent the real-world in a stylized and simplified way) or to approach them with inappropriate methods (which might lead to suboptimal and misleading results without being able to recognize these errors). This contribution is mainly concerned with the effects of market frictions on financial management which are therefore explicitly taken into account. Hence, for reliable results, an alternative class of optimization techniques has to be employed that are capable of dealing with these frictions, namely *heuristic optimization* techniques.

Opposed to the well-defined problems considered so far, there also exist problems where the underlying structure is unknown, partially hidden – or simply too complex to be modeled. When an underlying structure can be assumed or when there are pairs of input/output data, these questions can be approached, e.g., with *econometric*[1] or *Artificial Intelligence*[2] methods. In finance, time series analysis, pricing of complex securities, model selection problems, and artificial markets would be typical examples.[3] In this contribution, however, only well-defined optimization problems will be considered.

2.1.2 Techniques for Hard Optimization Problems

2.1.2.1 *Measuring Computational Complexity*

Before introducing specific optimization methods, it might be helpful to find a classification for the size and complexity of the considered problems – and, in due course, a measure of the methods applied on them. The computational complexity of an optimization problem as well as optimization procedures (and algorithms in general) is frequently given in $\mathcal{O}(\cdot)$ notation which indicates the asymptotic time necessary to solve the problem when it involves n (instances of the) decision variables and the problem size is determined by the number of these decision variables. An algorithm of order $\mathcal{O}(n)$, e.g., will consume a processor time (CPU time) of $n \cdot c$, i.e., the time necessary for solving the problem increases linearly in the number of instances;[4] a polynomial algorithm of order $\mathcal{O}(n^k)$ will consume $c \cdot n^k$ where k is

---

[1] See, e.g., Winker (2001) or Gourieroux and Jasiak (2001).

[2] See, e.g., Russell and Norvig (2003).

[3] See, e.g., Winker and Gilli (2004), Kontoghiorghes, Rustem, and Siokos (2002), Rasmussen, Goldy, and Solli (2002) or LeBaron (1999).

[4] Note that these considerations exclude aspects such as memory management or time consumed by interface communication. Practical implementations should account for these machine and programming environment depending characteristics.

a constant specific for the problem or algorithm. E.g., reading n pages of a book might consume linear time (i.e., $\mathcal{O}(n)$), but finding out whether there are any duplicates in a pile of n pages demands that each of them is compared to the remaining $(n-1)$ pages, and the complexity becomes $\mathcal{O}(n \cdot (n-1)) \approx \mathcal{O}(n^2)$ and is therefore quadratic in n.

The constant c will differ across programming platforms as well as actual CPU capacities. For sufficiently large n, the main contribution to computational time will come from the argument related to n or, if the argument consists of several components, the "worst" of them, i.e., the one that eventually outgrows the other: When the complexity is $\mathcal{O}\left(\ln(n) \cdot (n/k)^2\right)$ then, for any constant k and sufficiently large n, the quadratic term will outweigh the logarithmic term and the CPU time can be considered quadratic in n. A sole increase in CPU power which implies a reduction in c will therefore not have a sustainable effect on the (general) computability of a problem. A real improvement of the complexity can only be achieved when a better algorithm is found. This applies not only for the (sometimes called "easy") polynomial problems where the exponent k can become quite large, too; it is all the more true for a special class of optimization and search problems: For the group of *non-deterministic polynomial time* (*NP*) *complete* problems, there is no deterministic algorithm known that can find an exact solution within polynomial time.[5] This means that for deciding whether the found solution is optimal, the number of necessary steps is not a polynomial, but (at least) an exponential function of the problem size in the worst case. Two well-known problems of this group are the *Traveling Salesman Problem* (*TSP*)[6] and the *Knapsack Problem* (*KP*)[7]. One of the difficulties with optimization

---

[5] Actually, it can be shown that if one of these problems could be solved in polynomial time, this solution could be transferred and all of these problems could be solved in polynomial time.

[6] In the TSP, a salesperson has to find the shortest route for traveling to n different cities, usually without visiting one city twice. The problem can be modeled by a graph where nodes are the cities and the arcs are the distances between any two cities. Finding the shortest route corresponds then to finding the shortes path through this graph (or the shortest *Hamiltonian cycle*, if the salesperson takes a round trip where the tour starts and ends in the same city). If the graph is fully connected, i.e., there is a direct connection between any two cities, there are $n!$ alternative routes to choose from – which also characterizes the worst case computational complexity of this problem.

[7] In the KP, a tourist finds a number of precious stones that differ in size and in value per size unit. As the capacity of her knapsack is limited, the task of this 1/0 knapsack problem is to select stones such that the value of the contents is maximized. Fast decisions to this problem are possible only under rare circumstances. See Kellerer, Pferschy, and Pisinger (2004) and section 4.2.1.

problems is that the complexity for solving them is not always apparent: Proofs that an optimization problem belongs to a certain complexity class are therefore helpful when deciding which solution strategy ought to be considered.[8]

2.1.2.2 Brute-Force Methods

Analytic, closed-form solutions to optimization problems are desirable as they allow for exact and straightforward answers. The major advantage of these solutions is that they can be derived without explicit knowledge of the included parameters' values: The optimization has to be done only once, and the result is in a form by which, given the relevant (exogenous) parameters, the optimal values for the decision variables can immediately be determined.

If such solutions do not exist, then the problem has usually to be solved for each individual set of parameters. The approach that needs the least optimization skills would be *complete enumeration* where simply all possible (and valid) values for the decision variables are tested. This approach has some severe downsides: first and foremost, it is frequently time-consuming far beyond acceptability. Second, it demands the set of candidate solutions to be discrete and finite; if the decision variables are continuous (i.e., have infinitely many alternatives) then they have to be discretized, i.e., transformed into a countable number of alternatives – ensuring that the actual optimum is not excluded due to too large steps while keeping the resulting number of alternatives manageable.

In chapter 4, e.g., the problem will be to select k out of N assets and optimize their weights. The problem size can quickly get out of hand: there are $\binom{N}{k} = N! / ((N-k)! \cdot k!)$ alternative combinations for selecting k out of N assets without optimizing the weights; correspondingly, the complexity of an exhaustive search would be $\mathcal{O}\left(\binom{N}{k}\right)$. Selecting just 10 out of 100 assets comes with $\binom{100}{10} = 1.73 \times 10^{13}$ alternatives. For each of these alternatives, the optimal weights had to be found: When the weights of 10 assets may be either zero or multiples of 10% and short-sales are disallowed, the granularity of the weights is $g = 1/10\%$ which comes with $k^g = 10^{10}$ possible weight structures per alternative; the complexity is then increased to $\mathcal{O}\left(\binom{N}{k} \cdot k^g\right)$. Having

---

[8] See Knuth (1997) and Harel (1993).

a computer that is able to evaluate a million cases per second, complete enumeration would take 1.32×10^{11} years – which is approximately ten times the time since the Big Bang. When k is increased by just one additional asset from 10 to 11 (other things equal), the CPU time would increase to 233 times the time since the Big Bang; and if, in addition, the granularity would be increased to multiples of 5% (which would still be too rough for realistic applications), then the CPU time increased to more than 6 trillion times since the Big Bang.

Some of the problems dealt with in the following chapters have opportunity sets that are magnitudes larger; complete enumeration is therefore not a realistic alternative, nor would a sheer increase in computational power (by faster CPU's or having parallel computers) do the trick.

2.1.2.3 Traditional Numerical Methods and Algorithmic Approaches

Traditional numerical methods are usually based on iterative search algorithms that start with a (deterministic or arbitrary) solution which is iteratively improved according to some deterministic rule.[9] For financial optimization, methods from *Mathematical Programming* are often applied as these methods can manage problems where the constraints contain not only equalities, but also inequalities. Which type of method should and can be applied depends largely on the type of problem:[10]

Linear Programming will be applied when the optimization problem has a linear objective function and its constraints, too, are all linear (in-)equalities. The most popular method is the *Simplex Algorithm* where first the inequalities are transformed into equalities by adding *slack variables* and then including and excluding *base variables* until the optimum is found. Though its worst case computational complexity is exponential, it is found to work quite efficiently

---

[9] For a concise presentation of applications in economics, see Judd (1998).

[10] The following list of methods is far from exhaustive. For more details, see, e.g., Hillier and Lieberman (2003) for a concise introduction to Operations Research, and Stepan and Fischer (2001) for quantitative methods in Business Administration. Hillier and Lieberman (1995) presents methods in mathematical programming, a presentation of major algorithmic concepts can be found in Knuth (1997). Seydel (2002) and Brandimarte (2002) tackle several issues in computational finance and present suitable numerical methods.

for many instances. Some parametric models for portfolio selection therefore prefer linear risk measures (accepting that these risk measures have less desirable properties than the variance).

Quadratic and Concave Programming can be applied when the constraints are linear (in-)equalities, yet the objective function is quadratic. This is the case for the Markowitz model (1.7); how this can be done, will be presented in section 2.1.2.4. When the Kuhn-Tucker conditions hold,[11] a modified version of the Simplex Algorithm exists that is capable of solving these problems – the computational complexity of which, however, is also exponential in the worst case.

Dynamic Programming is a general concept rather than a strict algorithm and applies to problems that have, e.g., a temporal structure. For financial multi-temporal problems, the basic idea would be to split the problem into several sub-problems which are all myopic, i.e., have no temporal aspects when considered separately. First, the sub-problem for the last period, T, is solved. Next the optimal solution for last but one period, $T - 1$, is determined, that leads to the optimal solution for T, and so on until all sub-problems are solved.

Stochastic Programming is concerned with optimization problems where (some of the) data incorporated in the objective function are uncertain. Usual approaches include recourse, assumption of different scenarios and sensitivity analyses.

Other types of Mathematical Programming include non-linear programming, integer programming, binary programming and others. For some specimen types of problems, algorithms exist that (tend to) find good solutions. To approach the optimization problem at hand, it has to be brought into a structure for which the method is considered to work – which, for financial optimization problems, often comes with the introduction of strong restrictions or assumptions on the "allowed" values for the decision variables or constraints.

Greedy Algorithms always prefer the next one step that yields the maximum improvement but does not assess its consequences. Given a current (suboptimal) solution, a greedy algorithm would search for a modified solution within

---

[11] See, e.g., Chiang (1984, section 21.4).

a certain neighborhood and choose the "best" among them. This approach is sometimes called *hill-climbing*, referring to a mountaineer who will choose her every next step in a way that brings the utmost increase. As these algorithms are focused on the next step only, they get easily stuck when there are many local optima and the initial values are not chosen well. Hence, this approach demands smooth solution spaces and a monotonous objective function for good solutions and is related to the concept of gradient search.

Gradient Search can be performed when the objective function $f(x)$ is differentiable and strictly convex[12] and the optimum can be found with the first order condition $\partial f / \partial x = 0$. Given the current candidate solution, the gradient $\nabla f(x') = \left(\frac{\partial f}{\partial x_1}, \dots, \frac{\partial f}{\partial x_n} \right)$ is computed for $x' = x$. The solution is readjusted according to $x' = x' + \delta \cdot \nabla f(x)$ which corresponds to $x'_j := x'_j + \delta \cdot \frac{\partial f}{\partial x_i} \Big|_{x=x'} \forall j$. This readjustment is repeated until the optimum x^* with $\nabla f(x) = 0$ is reached. Graphically speaking, this procedure determines the tangency at point x' and moves the decision variables towards values for which the tangency's slope is expected to be 0 and any slight change of any x_j would worsen the value of the objective function f.

Divide and Conquer Algorithms iteratively split the problem into sub-problems until the sub-problems can be solved in reasonable time. These partial results are then merged for the solution of the complete problem. These approaches demand that the original problem can be partitioned in a way that the quality of the solutions for sub-problems will not interfere with each other, i.e., that the sub-problems are not interdependent.

Branch and Bound Algorithms can be employed in some instances where parts of the opportunity set and candidate solutions can be excluded by selection tests. The idea is to iteratively split the opportunity space into subsets and identify as soon as possible those subsets where the optimum is definitely not a member of, mainly by either repeatedly narrowing the boundaries within which the solution must fall, by excluding infeasible solutions, or by "pruning" those solutions that are already outperformed by some other solution

---

[12] Here, maximization problems are considered. For minimization problems, the similar arguments for concave functions can be considered. Note that any maximization problem can be transformed into a minimization problem (usually by taking the inverse of the objective function or multiplying it by -1) and *vice versa*.

found so far. The opportunity set is therefore repeatedly narrowed down until either a single valid solution is found or until the problem is manageable with other methods, such as complete enumeration of all remaining solutions or another numerical method.

As mentioned earlier, a salient characteristic of these methods is that they work only for problems which satisfy certain conditions: The objective function must be of a certain type, the constraints must be expressible in certain formats, and so forth. Their application is therefore restricted to a rather limited set of problems. In practice, these limitations are often circumvent by modifying the problems and stating the problems is a way that they are solvable. Another main caveat of these optimization methods is that they are mostly based on rather strict deterministic rules. Hence, they might produce wrong solutions when the considered problem has not just one global, but also one or several local optima. Once deterministic search rules converge towards such local optima, they might have problems leaving them again (and therefore will never find the global optimum), given they converge in the first place. Also, deterministic rules have it that, by definition, for a given situation, there is a unique response. A deterministic search algorithm will therefore always produce the same result for a given problem when the search strategy cannot be influenced and the initial values, too, are chosen deterministically. This being the standard case, repeated runs will always report the same local optimum, in particular when the initial value for the search process is found with some deterministic rule, too.

In the lack of alternatives, however, financial optimization problems have often been modeled in a way that they can be solved with one of these methods. As a consequence, they either had to be rather restrictive or had to accept strong simplifications (such as the assumption of frictionless markets in order to satisfy the Kuhn-Tucker conditions), or accepted that the solutions are likely to be suboptimal (such as, e.g., in *Asset Class Management*, where the universe of available assets is split into submarkets (subportfolios) which are optimized independently in a "divide and conquer" fashion, ignoring the relationships between the classes). However, without this fitting of the problems to the available methods, the majority of (theoretical and practical) optimization problems in portfolio management could not readily be answered. Due to the fitting, on the other hand, it is difficult (and quite often impossible) to tell whether a reported solution is unique or just one out of many optima and how far away this reported solution is from the global optimum.

2.1.2.4   Portfolio Optimization with Linear and Quadratic Programming

Many software packages for optimization problems offer routines and standard solutions for Linear and Quadratic Programming Problems. Linear Programming (LP)
can be applied when the objective function is linear in the decision variable and the
constraints are all equalities or inequalities that, too, are linear. A general statement
would be

$$\min_x f'x$$

subject to

$$Ax = a$$
$$Bx \leq b$$

where x is the vector of decision variables and A, B, a, and b are matrices and vectors, respectively, that capture the constraints. Note that any minimization problem
can be transformed into a maximization problem simply by changing the sign of the
objective function (i.e., by multiplying f with -1) and that by choosing the appropriate signs, inequalities of the type $Cx \geq c$ can be turned into $-Cx \leq -c$, i.e., b
can contain upper and lower limits alike.

A simple application in portfolio selection might be to find the weights $x_i, i =
1, \ldots, N$, that maximize the portfolio's expected return when the weight of each of the
N assets must be within a given range, i.e., $x^{\ell} \leq x_i \leq x^u$, and the weights must add
up to one, i.e., $\sum_i x_i = 1$. This can be achieved by setting $f = -r$; $A = 1_{1 \times N}$, $a = 1$,
$B = [-I_{N \times N} \quad I_{N \times N}]'$ and $b = [-x^{\ell} 1_{1 \times N} \quad x^u 1_{1 \times N}]'$ where r is the vector of expected
returns and I and 1 are the respective identity matrices and unity vectors with the
dimensions as indexed. This approach, however, is not able to cope with variance or
volatility as these are quadratic risk measures.

Quadratic Programming (QP) problems, like LP problems, have only constraints
that can be expressed as linear (in-)equalities with respect to the decision variables;
their objective function, however, allows for an additional term that is quadratic in
the decision variables. A standard general statement therefore might read as follows:

$$\min_x f'x + \frac{1}{2}x'Hx$$

subject to

$$Ax = a$$
$$Bx \leq b.$$

This can be applied to determine the Markowitz efficient portfolio for a return of r_P by implementing the model (1.7) as follows. If r and Σ denote the return vector and covariance matrix, respectively, then $f = 0_{1 \times N}$, $H = 2\Sigma$, $A = [1_{N \times 1} \quad r]'$, $a = [1 \quad r_P]'$, $B = -I_{N \times N}$ and $b = 0_{N \times 1}$ where 0 is the zero vector.

If, however, the whole efficient line of the Markowitz model is to be identified, then the objective function (1.7a*) is to be applied and the respective parameters are $f = -\lambda r$, $H = 2(1 + \lambda)\Sigma$, $A = 1_{1 \times N}$, $a = 1$, $B = -I_{N \times N}$ and $b = 0_{N \times 1}$. Since λ measures the trade-off between risk and return, $\lambda = 0$ will lead to the identification of the Minimum Variance Portfolio. On the other hand, $\lambda = 1$ puts all the weight on the expected return and will therefore report the portfolio with the highest possible yield which, for the given model with non-negativity constraints but no upper limits on x_i, will contain exclusively the one asset with the highest expected return. To identify the efficient portfolios between these two extremes, a usual way would be to increase λ in sufficiently small steps from zero to one and solve the optimization problem for these values.

In a Tobin framework as presented in section 1.1.2.3 where the set of risky assets is supplemented with one safe asset, the investor will be best off when investing an amount α into the safe asset and the remainder of $(1 - \alpha)$ into the tangency portfolio $\mathcal{T}$. Given an exogenously chosen value $0 < \alpha < 1$ (for convenience, $\alpha \to 0$), the respective parameters for the quadratic programming model are $f = -[r' \quad r_s]'$, $H = 2 \begin{bmatrix} \Sigma & 0_{N \times 1} \\ 0_{1 \times N} & 0 \end{bmatrix}$, $A = \begin{bmatrix} 1_{1 \times N} & 1 \\ 0_{1 \times N} & 1 \end{bmatrix}$, $a = [1 \quad \alpha]'$, $B = -I_{(N+1) \times (N+1)}$ and $b = 0_{(N+1) \times 1}$. The resulting vector x is of dimension $(N + 1) \times 1$, where the first N elements represent $(1 - \alpha) \cdot x_{\mathcal{T}}$, whereas the $(N + 1)$-st element is the weight of the safe asset in the investor's overall portfolio and (by constraint) has the value of α. By the separation theorem, the weights for $\mathcal{T}$ can then be determined by $x_{\mathcal{T}} = \frac{1}{1 - \alpha} \begin{bmatrix} x_1 & \cdots & x_N \end{bmatrix}'$.

2.1.2.5   *"Traditional" Deterministic versus Stochastic and Heuristic Methods*

Classical optimization techniques as presented so far can be divided into two main groups. The first group of methods is based on exhaustive search or (complete) enumeration, i.e., testing all candidate solutions. The crux of approaches like branch and bound is to truncate as much of the search space as possible and hence to eliminate groups of candidates that can been identified as inferior beforehand. However, even after pruning the search space, the remaining number of candidates might still exceed the available capacities, provided the number of solutions is discrete and finite in the first place.

The second type comprises techniques that are typically based on the differential calculus, i.e., they apply the first order conditions and push the decision variables towards values where the first derivative or gradient of the objective function is (presumably) zero. An implicit assumption is that there is just one optimum and/or that the optimum can be reached on a "direct path" from the starting point. The search process itself is usually based on deterministic numerical rules. This implies that, given the same initial values, repeated runs will always report the same result – which, as argued, is not necessarily a good thing: repeated runs with same (deterministically generated) initial values will report the same results, unable to judge whether the global or just a local optimum has been found. To illustrate this problem, reconsider the function depicted in Figure 2.1 on page 39. If the initial guess is a value for x that is near one of the local optima x_A or x_C, then a traditional numerical procedure is likely to end up at the local maximum closest to the initial guess, and the global optimum, x_B, will remain undiscovered. In practice, the deterministic behavior and the straightforward quest for the closest optimum from the current solutions perspective can be a serious problem, in particular when there are many local optima which are "far apart" from the global optimum, but close to the starting value. Also, slight improvements in the objective function might come with substantially different values for the decision variables.

One radical alternative to deterministic methods would be *Monte Carlo (MC) search*: A large number of random (yet valid with respect to the constraints) guesses for values for the decision variables are generated and the respective values of the objective function are determined.[13] With a sufficiently large number of independent guesses, this approach is likely to eventually identify the optimum or at least to identify regions within which it is likely or unlikely to be found. This concept is much more flexible than numerical methods as its main restrictions are *a priori* the availability of a suitable random number generator and the time necessary to perform a sufficiently large number of tries. It can therefore be applied to narrow down the search space which could then be approached with numerical methods. The major downside of it is, however, that it might be quite inefficient and inexact: Quite often, significant parts of the opportunity set can quickly be identified as far from the actual optimum; further search in this "region" is therefore just time consuming.

Heuristic search methods and *heuristic optimization techniques* also incorporate stochastic elements. Unlike Monte Carlo search, however, they have mechanisms that drive the search towards promising regions of the opportunity space. They therefore combine the advantages of the previously presented approaches: much like numerical methods, they aim to converge to the optimum in course of iterated search, yet they are less likely to end up in a local optimum and, above all, are very flexible and therefore are less restricted (or even perfectly unrestricted) to certain forms of constraints.

The heuristics discussed in due course and applied in the main part of this contribution were designed to solve optimization problems by repeatedly generating and testing new solutions. These techniques therefore address problems where there actually exist a well-defined model and objective function. If this is not the case, there exist alternative methods in *soft computing*[14] and *computational intelligence*[15].

---

[13] It is not always possible to guarantee beforehand that none of the constraints is violated; also, ascertaining that only valid candidate solutions are generated might be computationally costly. In these cases, a simple measure would be not to care about these constraints when generating the candidate solutions but to add a *punishment term* to the objective value when this candidate turns out to be invalid.

[14] Coined by the inventor of *fuzzy logic*, Lotfi A. Zadeh, the term *soft computing* refers to methods and procedures that not only tolerate uncertainty, fuzziness, imprecision and partial correctness but also make use of them; see, e.g., Zadeh and Garibaldi (2004).

[15] Introduced by James Bezdek, *computational intelligence* refers to methods that use numerical procedures to simulate intelligent behavior; see Bezdek (1992, 1994).

A popular method of this type are *Neural Networks* which mimic the natural brain process while learning by a non-linear regression of input-output data.[16]

2.2   Heuristic Optimization Techniques

2.2.1   Underlying Concepts

The toy manufacturer Hasbro, Inc., produces a popular game called *Mastermind*. The rules of this game are rather simple: one player selects four colored pegs and the other player has to guess their color and sequence within a limited number of trials. After each guess, the second player is told how many of the guessed pegs are of the right color and how many are the right color and in the right position. The problem is therefore well-defined, as there are a clear objective function and a well-defined underlying model: though the "parameters" of the latter are hidden to the second player, it produces a uniquely defined feedback for any possible guess within a game.

Although there are 360 different combinations in the standard case[17] the second player is supposed to find the right solution within eight guesses or less. Complete enumeration is therefore not possible. The typical beginner's approach is to perform a Monte Carlo search by trying several perfectly random guesses (or the other player's favorite colors) and hoping to find the solution either by sheer chance or by eventually interpreting the outcome of the independent guesses. With unlimited guesses, this strategy will eventually find the solution; when limited to just eight guesses, the hit rate is disappointingly low.

More advanced players also start off with a perfectly random guess, but they reduce the "degree of randomness" in the subsequent guesses by considering the outcomes from the previous guesses: E.g., when the previous guess brought two white

---

[16] See Russell and Norvig (2003) for a general presentation; applications to time series forecasting are presented in Azoff (1994).

[17] The standard case demands all four pegs to be of different color with six colors to choose from. Alternative versions allow for "holes" in the structure, repeated colors and/or also the "white" and "black" pegs, used to indicate "correct color" and "correct color and position", respectively – resulting in up to 6 561 combinations.

pegs (i.e., only two right colors, none in the right position, and two wrong colors), the next guess should contain some variation in the color; if the answer were four white pegs (i.e., all the colors are right, yet all in the wrong position), the player can concentrate on the order of the previously used pegs rather than experimenting with new colors. The individual guesses are therefore not necessarily independent, yet (usually) there is no deterministic rule for how to make the next guess. *Mastermind* might therefore serve as an example where the solution to a problem can be found quite efficiently by applying an appropriate *heuristic optimization method*.

2.2.2   Characteristics of Heuristic Optimization Methods

The central common feature of all *heuristic optimization* (*HO*) methods is that they start off with a more or less arbitrary initial solution, iteratively produce new solutions by some generation rule and evaluate these new solutions, and eventually report the best solution found during the search process. The execution of the iterated search procedure is usually halted when there has been no further improvement over a given number of iterations (or further improvements cannot be expected); when the found solution is good enough; when the allowed CPU time (or other external limit) has been reached; or when some internal parameter terminates the algorithm's execution. Another obvious halting condition would be exhaustion of valid candidate solutions – a case hardly ever realized in practice.

Since HO methods may differ substantially in their underlying concepts, a general classification scheme is difficult to find. Nonetheless, the following list highlights some central aspects that allow for comparisons between the methods.[18] With the rapidly increasing number of new heuristics and variants or combinations of already existing ones, the following list and the examples given therein are far from exhaustive.

Generation of new solutions.   A new solution can be generated by modifying the current solution (*neighborhood search*) or by building a new solution based on past experience or results. In doing so, a deterministic rule, a random guess or a combination of both (e.g., deterministically generating a number of alternatives and randomly selecting one of them) can be employed.

---

[18]   For an alternative classification, see, e.g., Silver (2002) and Winker and Gilli (2004).

Treatment of new solutions. In order to overcome local optima, HO methods usually consider not only those new solutions that lead to an immediate improvement, but also some of those that are knowingly inferior to the best solution found so far. To enforce convergence, however, inferior solutions might either be included only when not being too far from the known optimum or might be given a smaller "weight." Also, the best found solution so far might be reinforced (*elitist principle*), new solutions might be ranked and only the best of them are kept for future consideration, etc. The underlying acceptance rules can be deterministic or contain certain randomness.

Number of search agents. Whereas in some methods, a single agent aims to improve her solution, population based methods often make use of collective knowledge gathered in past iterations.

Limitations of the search space. Given the usually vast search space, new solutions can be found by searching within a certain neighborhood of a search agent's current solution or of what the population (implicitly) considers promising. Some methods, on the other hand explicitly exclude certain neighborhoods or regions to avoid cyclic search paths or spending too much computation time on supposedly irrelevant alternatives.

Prior knowledge. When there exist general guidelines of what is likely to make a good solution, this prior knowledge can be incorporated in the choice of the initial solutions or in the search process (*guided search*). Though the inclusion of prior knowledge might significantly reduce the search space and increase the convergence speed, it might also lead to inferior solutions as the search might get guided in the wrong direction or the algorithm might have severe problems in overcoming local optima. Prior knowledge is therefore found in a rather limited number of HO methods and there, too, rather an option than a prerequisite.

Flexibility for specific constraints. Whereas there exist true general purpose methods that can be applied to virtually any type of optimization problem, some methods are tailor-made to particular types of constraints and are therefore difficult to apply to other classes of optimization problems.

Other aspects allow for testing and ranking different algorithms and might also affect the decision which method to select for a particular optimization problem:

Ease of implementation. The (in-)flexibility of the concept, the complexity of the necessary steps within an iteration step, the number of parameters and the time necessary to find appropriate values for these parameters are a common first selection criterion.

Computational complexity. For HO methods the complexity depends merely depends on the costs for evaluating per candidate solution, on the number of iterations, and, if applicable, on the population size and the costs of administrating the population. Though the number of iterations (and population's size) will usually increase for larger problem spaces, the resulting increase in computational costs is usually substantially lower than it would be for traditional methods. Hence, the computational complexity of HO methods is comparatively low; even for NP complete problems, many HO algorithms have at most polynomial complexity. General statements, however, are difficult to make due to the differences in the HO techniques and, above all, the differences in the optimization problems' complexities.

Convergence speed. The CPU time (or, alternatively, the number of evaluated candidate solutions) until no further improvement is found, is often used as a measure to compare different algorithms. Speed might be a salient property of an algorithm in practical solutions – though not too meaningful when taken as a sole criterion as it does not necessarily differentiate between local and global optimum convergence and as long as a "reasonable" time limit is not exceeded.

Reliability. For some major heuristics, proofs exist that these methods will converge towards the global optimum – given sufficient computation time and an appropriate choice of parameters. In practice, one often has to accept a trade-off between low computational time (or high convergence speed) and the chance that the global optimum is missed. With the inherent danger of getting stuck in a local optimum, heuristics are therefore frequently judged by their ratio of reporting local optima or other inferior solutions.

To reduce the vagueness of these aspects, section 2.3 presents some of the major and commonly used heuristics that are typical representatives for this type of methods and that underline the differences in the methods with regard to the above mentioned aspects: In *Threshold Accepting* (section 2.3.1), one solution is considered at a time and iteratively modified until it reaches an optimum; in *Evolution Based Methods* (section 2.3.2), a number of promising solutions are further evolved at the same time; in *Ant Systems* (section 2.3.3), collective experience is used; and *Memetic Algorithms* (section 2.3.4) are a typical example for successful hybrid algorithms where the advantages of several methods could be combined.[19] There is neither a heuristic that outperforms all other heuristics whatever the optimization problem, nor can one provide a general implementation scheme regardless of the problem type.[20] The presentation in this introductory chapter is therefore reduced to the fundamental underlying ideas of the heuristics; more detailed descriptions will be offered when applied in the subsequent chapters.

2.3 Some Selected Methods

2.3.1 Simulated Annealing and Threshold Accepting

Kirkpatrick, Gelatt, and Vecchi (1983) present one of the simplest and most general HO techniques which turned out to be one of the most efficient ones, too: *Simulated Annealing (SA)*. This algorithm mimics the crystallization process during cooling or annealing: When the material is hot, the particles have high kinetic energy and move more or less randomly regardless of their and the other particles' positions. The cooler the material gets, however, the more the particles are "torn" towards the direction that minimizes the energy balance. The SA algorithm does the same when searching for the optimal values for the decision parameters: It repeatedly suggests random modifications to the current solution, but progressively keeps only those that improve the current situation.

SA applies a probabilistic rule to decide whether the new solution replaces the current one or not. This rule considers the change in the objective function (mea-

---

[19] For general presentations and comparisons of HO methods, see, e.g., Osman and Kelly (1996), Taillard, Gambardella, Gendreau, and Potvin (2001), Michalewicz and Fogel (1999), Aarts and Lenstra (2003) or Winker and Gilli (2004). Osman and Laporte (1996) offer an extensive bibliography of the theory and application of meta-heuristics, including 1 380 references. Ausiello and Protasi (1995) investigate local search heuristics with respect to NP optimization problems.

[20] See, e.g., Hertz and Widmer (2003).

```
generate random valid solution x;
REPEAT
    generate new solution x' by randomly modifying
        the current solution x;
    evaluate new solution x';
    IF acceptance criterion is met THEN;
        replace x with x';
    END;
    adjust acceptance criterion;
UNTIL halting criterion is met;
```

Listing 2.1: Basic structure for Simulated Annealing (SA) and Threshold Accepting (TA)

suring the improvement/impairment) and an equivalent to "temperature" (reflecting the progress in the iterations). Dueck and Scheuer (1990) suggest a deterministic acceptance rule instead which makes the algorithm even simpler: Accept any random modification unless the resulting impairment exceeds a certain threshold; this threshold is lowered over the iterations. This algorithm is known as *Threshold Accepting (TA).*

Listing 2.1 summarizes the pseudo-code for SA and TA where the values for the elements of a vector *x* are to be optimized; SA will be presented in more details when applied in chapter 3; different acceptance criteria will be compared in section 6.3.1. Both SA and TA usually start off with a random solution and generate new solutions by perfectly random search within the current solution's neighborhood. In either method, the acceptance of impairments allows to overcome local optima. To avoid a Monte Carlo search path, however, improvements are more likely to be accepted than impairments at any stage, and with decreasing tolerance on impairments, the search strategy shifts towards a hill-climbing search.

SA and TA are both extremely flexible methods which are rather easy to implement. Both are general purpose approaches which cause relatively little computational complexity and for which convergence proofs exist.[21] Single agent neighborhood search methods such as SA and TA have proofed successful when the solution space is not too rough, i.e., if the number of local optima is not too large.[22]

---

[21] See Aarts and van Laarhoven (1985) and Althöfer and Koschnik (1991), respectively.

[22] For a concise presentation of TA, its properties and applications in economics as well as issues related to evaluating heuristically obtained results, see Winker (2001).

```
generate P random solutions x₁ ... xₚ;
REPEAT
    FOR each parent individual i = 1...P
        generate offspring x'ᵢ by randomly modifying
            the "parent" xᵢ;
        evaluate new solution x'ᵢ;
    END;
    rank parents and offspring;
    select the best P of these solutions for new parent population;
UNTIL halting criterion met;
```

Listing 2.2: Basic structure for Evolutionary Strategies (ES)

2.3.2 Evolution Based and Genetic Methods

Inspired by their natural equivalent, the ideas of simulated evolution and artificial life have gained some tradition in machine learning and, eventually, in heuristic optimization.[23] One of the first algorithms actually addressing an optimization problem are *Evolutionary Strategies* (*ES*) by Rechenberg (1965). Here, a population of P initial solution vectors is generated. In each of the following iteration steps, each individual is treated as a parent that produces one offspring by adding a random modification to the parent's solution. From the now doubled population, only the best P agents are selected which will constitute the parent population in the next generation. Listing 2.2 summarizes the main steps of this original concept. Later versions offer modifications and improvements; in Rechenberg (1973), e.g., multiple parents generate a single offspring.

Evolution based methods gained significant recognition with the advent of *Genetic Algorithms* (*GA*). Based on some of his earlier writings as well as related approaches in the literature, Holland (1975) attributes probabilities for reproduction to the individual "chromosomes," x_i, that reflect their relative fitness within the population. In the sense of the "survival of the fittest" principle, high fitness increases the chances of (multiple) reproduction, low fitness will ultimately lead to extinction. New offspring is generated by combining the chromosomes of two parent chromosomes; in the simplest case this cross-over can be done by "cutting" each parent's chromosomes into two pieces and creating two siblings by recombining each par-

---

[23] A survey on these topics can be found in Fogel (1998).

```
generate P random chromosomes;
REPEAT
    determine fitness of all chromosomes i = 1...P;
    determine replication probabilities pᵢ based on relative fitness;
    FOR number of reproductions;
        randomly select two parents based on pᵢ;
        generate two children by cross-over operation on parents;
    END;
    insert offspring into the population;
    remove P chromosomes based on inverse replication probability;
    apply mutation to some/all individuals;
UNTIL halting criterion met;
```

Listing 2.3: Basic structure for Genetic Algorithms (GA)

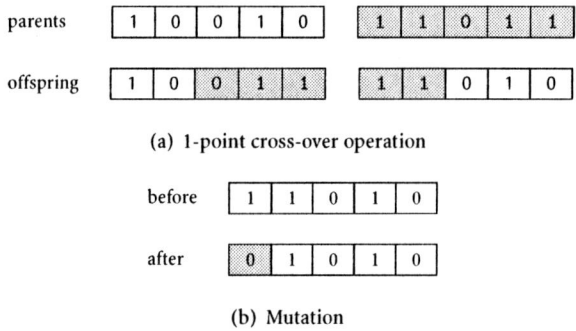

(a) 1-point cross-over operation

(b) Mutation

Fig. 2.2: Examples for evolutionary operators on binary strings

ent's first part with the other parent's second part (see Figure 2.2(a)). In addition mutation can take place, again by randomly modifying an existing (i.e., parent's or newly generated offspring's) solution (see Figure 2.2(b)).

Over the last decades, GA have become the prime method for evolutionary optimization – with a number of suggestions for alternative cross-over operations (not least because GA were originally designed for chromosomes coded as binary strings), mutation frequency, cloning (i.e., unchanged replication) of existing chromosomes, etc. Listing 2.3 therefore indicates just the main steps of a GA; the structure of actual implementations might differ. Fogel (2001) offers a concise overview of methods and literature in evolutionary computation.

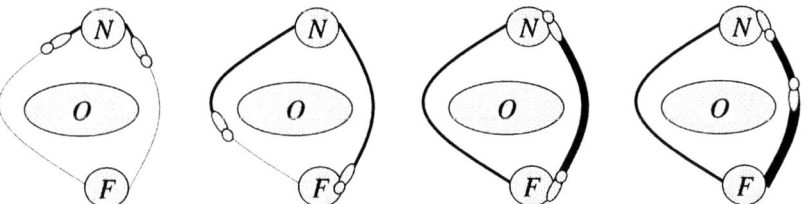

Fig. 2.3: Simple foraging example for a colony with two ants

Whereas SA and TA are single-agent methods where a solution is persistently modified (or "mutated"), evolutionary methods have to administer whole populations. At the same time, they all derive their new solutions by modifying existing current solutions. Evolutionary methods are more demanding to implement than are SA and TA. Also, they are more time-consuming because of their computational costs for administrating the population. At the same time, they are less likely to get stuck in local optima as the respective chromosomes are likely to be eventually be replaced with "fitter" alternatives.

2.3.3   Ant Systems and Ant Colony Optimization

Evolution has provided ants with a simple, yet enormously efficient method of finding shortest paths.[24] While traveling, ants lay pheromone trails which help themselves and their followers to orientate.

To illustrate the underlying principle, we assume a nest N and a food source F are separated by an obstacle O (Figure 2.3) and that there are two alternative routes leaving N both leading to F, yet different in length. Since the colony has no information which of the two routes to choose, the population (here consisting of two ants) is likely to split up and each ant selects a different trail. Since the route on the right is shorter, the ant on it reaches F while the other ant is still on its way. Supplied with food, the ant wants to return to the nest and finds a pheromone trail (namely its own) on one of the two possible ways back and will therefore select this alternative with a higher probability. If it actually chooses this route, it lays a second

---

[24] See Goss, Aron, Deneubourg, and Pasteels (1989).

pheromone trail while returning to the nest. Meanwhile the second ant has reached F and wants to bring the food to the nest. Again, F can be left on two routes: the left one (=long) has now one trail on it, the right one (=short) has already two trails. As the ant prefers routes with more pheromone in it, it is likely to return on the right path – which is the shorter one and will then have a third trail on it (versus one on the left path). The next time the ants leave the nest, they already consider the right route to be more attractive and are likely to select it over the left one. In real live, this self-reinforcing principle is enhanced by two further effects: shorter routes get more pheromone trails as ants can travel on them more often within the same time span than they could on longer routes; and old pheromone trails tend to evaporate making routes without new trails less attractive.

Based on this reinforcement mechanism, the tendency towards the shorter route will increase. At the same time, there remains a certain probability that routes with less scent will be chosen; this assures that new, yet unexplored alternatives can be considered. If these new alternatives turn out to be shorter (e.g., because to a closer food source), the ant principle will enforce it, and – on the long run – it will become the new most attractive route; if it is longer, the detour is unlikely to have a lasting impression on the colony's behavior.

Dorigo, Maniezzo, and Colorni (1991) transfer this metaphor to heuristic optimization called *Ant System* (*AS*) by having a population of artificial ants search in a graph where the knots correspond to locations and the arcs represent the amount of pheromone, i.e., attractiveness of choosing the path linking these locations. Being placed at an arbitrary location and having to decide where to move next, the artificial ant will choose (among the feasible) routes those with higher probability that are marked with more pheromone. The pheromone is usually administered in a pheromone matrix where two basic kinds of updates take place: on the one hand, new trails are added that are the stronger the more often they are chosen and the better the corresponding result; on the other hand, trails evaporate making rarely chosen paths even less attractive.

Since the original concept of AS parallels the Traveling Salesman Problem,[25] Listing 2.4 presents this algorithm for the task of finding the shortest route when a given number of cities have to be visited. Meanwhile, there exist several extensions and

---

[25] See section 2.1.2.1.

```
initialize trails and parameters;
REPEAT
    FOR all ants do;
        deposit ant at a random location;
        REPEAT
            select randomly next city according to pheromone trail;
        UNTIL route complete;
        determine tour length;
    END;
    let a fixed proportion of all pheromone trails evaporate;
    FOR all ants DO;
        add pheromone to chosen paths (more for shorter tours);
    END;
UNTIL halting criterion met;
```

Listing 2.4: Basic Structure for Ant System (AS)

variants most of which suggest improved trail update rules or selection procedures leading to higher reliability. Also there exist modifications to open this algorithm for optimization problems other than ordering. A survey can be found in Bonabeau, Dorigo, and Theraulaz (1999).

The concept of the pheromone matrix facilitates the gathering and sharing of collective knowledge and experience: While in the previously presented methods SA, TA and GA derive their new solutions from one (or two paternal) existing solution(s) and adding a random term to it, the contributions of many ants (from the current and past generations) support the generation of a new solution. As a result, ant based systems usually have high convergence speed and reliability – yet are also computationally more demanding as trail updates and the generation of new solutions is more complex. Another disadvantage is that ant based algorithms are less flexible in their application.

2.3.4   Memetic Algorithms

Single agent neighborhood search methods such as SA or TA, where one solution is modified step by step until convergence, are successful in particular when there is a limited number of local optima, when the agent can at least roughly figure out in which direction the global optimum can be expected and when this optimum

```
initialize population;
REPEAT
    perform individual neighborhood search;
    compete;
    perform individual neighborhood search;
    cooperate;
    adjust acceptance criterion;
UNTIL halting criterion met;
```

Listing 2.5: Basic Structure for a Memetic Algorithm (MA)

is easily reachable given the step size and the distance between initial and optimal solution. If the algorithm appears to have problems of finding a solution or is likely to get stuck in local optima, one common remedy is to have a higher number of independent runs with different starting points, i.e., the optimization problem is solved repeatedly, and eventually the best of all found solutions is reported. Though the advantage of the independence between the runs is that mislead paths to local optima cannot be misleading in the current search, prior experience is lost and has to be gathered again. This increases inefficiency and run time. In population based methods such as GA, a whole population of agents produces several solutions at a time, which are regularly compared and the best of which are combined or re-used for new solutions. Population based methods therefore tend to be more likely to (eventually) overcome local optima. At the same time, they might have problems when already being close to the optimum where local neighborhood search would easily do the trick.

Moscato (1989) therefore suggests a method that combines the advantages of both concepts by having a population of agents that individually perform local search in a SA like fashion. In addition to the agents' independent neighborhood searches, they also compete and cooperate: competition is done in a tournament fashion where one agent challenges another and, if winning, imposes his solution onto the challenged agent; cooperation can be achieved by combining solutions with a cross-over operation as known, e.g., from GA. Unlike in other evolutionary methods, however, replacement in competition and cooperation uses the SA acceptance criterion instead of the replication probabilities and is therefore less time consuming. Listing 2.5 indicates the main steps of a simple version of MA.

This algorithm was inspired by a concept of Oxford zoologist Richard Dawkins who found that ideas and cultural units sometimes behave like "selfish" genes: they might be passed on from one person to another, they might be combined with other ideas, they mutate over time, and they have a tendency to self-replication. To resemble these properties, Dawkins introduced the term *meme* that reflects the French word for "self," *même*, and is pronounced in a way that it rhymes with "gene."[26]

MA as presented here[27] is a typical *hybrid algorithm* that combines elements of other algorithms and enhances them with original ideas and approaches. Compared to the other algorithms presented so far, MA has lower computational complexity than GA (yet, of course, higher complexity than a pure SA implementation). Being more flexible in shifting between independent neighborhood search and joint population search, they are more flexible than the methods they are built on.

2.4   Heuristic Optimization at Work

2.4.1   Estimating the Parameters for GARCH Models

2.4.1.1   *The Estimation Problem*

In section 1.1.3, different ways for estimating the volatility were presented, including GARCH models where the volatility can change over time and is assumed to follow an autoregressive process. Applying these models, however, is not always trivial as the parameters have to be estimated by maximizing the likelihood function (1.14) (see page 22) which might have many local optima. In the lack of closed-form solutions, traditional numerical procedures are usually employed – which might produce quite different results.

---

[26] See Dawkins (1976, chapter 7). For a more in-depth presentation and discussion of the *meme* concept and its application in social sciences, see, e.g., Blackmore (1999).

[27] Meanwhile, the literature holds many different versions of Memetic Algorithms, some of which are population based whereas others aren't, where the local search is not based on SA but on alternative methods such as Fred Glover's *Tabu Search* (where a list of recently visited solutions is kept that must not revisited again in order to avoid cycles), etc.; more details can be found in several contributions in Corne, Glover, and Dorigo (1999).

Given a time series r_t, Fiorentini, Calzolari, and Panattoni (1996) consider the simple GARCH(1,1) model (our notation)

$$r_t = \mu - e_t, \quad e_t | \Omega_{t-1} \sim N(0, \sigma_t^2) \tag{2.1a}$$

$$\sigma_t^2 = \alpha_0 + \alpha_1 \cdot e_{t-1}^2 + \beta_1 \cdot \sigma_{t-1}^2 \tag{2.1b}$$

with the objective of maximizing the conditional likelihood function, apart from the constant $-T/2 \cdot \ln(2 \cdot \pi)$,

$$\max_{\psi} \mathscr{L}(\psi) = \sum_{t=1}^{T} \left(-\frac{1}{2} \ln(\sigma_t^2) - \frac{1}{2} \frac{e_t^2}{\sigma_t^2} \right) \tag{2.1c}$$

where $\psi = [\mu, \alpha_0, \alpha_1, \beta_1]$ is the vector of decision variables and Ω_{t-1} is the information set available at time $t - 1$. They present a closed-form analytical expressions for the second derivatives of (2.1c) which can be used for initial values of ψ; for the actual search, they test gradient methods.

Based on these results, Bollerslev and Ghysels (1996) provide parameter estimations for the daily German mark/British pound exchange rate.[28] Their estimates for the coefficients are then used for benchmarks by Brooks, Burke, and Persand (2001) who estimate the parameters for the same data set with nine different specialized software packages. They find that only one of these packages is able to hit the benchmark coefficients and Hessian-based standard errors using the default settings. As this data set has become a benchmark problem for GARCH estimation[29] it can be used as a first example to illustrate how a heuristic optimization algorithm might be implemented and how the algorithm's performance can be optimized.

2.4.1.2   A Simple Heuristic Approach

To illustrate how to use heuristic optimization techniques and to test whether the obtained results are reliable, we approach the maximization problem (2.1) with one of the simpler of the introduced HO techniques, namely Simulated Annealing (SA) which has been introduced in section 2.3.1. Based on the pseudo-code in listing 2.1, the SA heuristic includes the following steps:

---

[28] The data, comprising 1974 observations, are available at
www.amstat.org/publications/jbes/ftp.html → viewing existing
publications → JBES APR-96 Issue → bollerslev.sec41.dat.

[29] See also McCullough and Renfro (1999).

- First, initial values for all decision variables (collected in the vector $\psi = [\mu, \alpha_0, \alpha_1, \beta_1]$) are generated by random guesses. The only prerequisite for these guesses is that the guessed values are "valid" with respect to the constraints.

- The main part of SA consists of a series of iterations where the following steps will be repeated:

  - The algorithm produces a new candidate solution, ψ', by modifying the current solution, ψ. To achieve this, one element j from ψ is selected arbitrarily. Then its current value is changed randomly. Formally, $\psi'_j = \psi_j + u \cdot \tilde{z}$ where $\tilde{z} \in [-1, +1]$ is an equally distributed random number. The other elements of ψ are left unchanged, i.e., $\psi'_k = \psi_k \; \forall k \neq j$.

  - Having generated a new candidate solution, ψ', the change in the objective function (here: the log-likelihood function) is calculated: $\Delta \mathscr{L} = \mathscr{L}(\psi') - \mathscr{L}(\psi)$. According to the SA principle, a stochastic acceptance criterion for the new solution is applied that takes the change in the objective function, $\Delta \mathscr{L}$, into account as well as how progressed the algorithm is: In early iterations, even large impairments have a considerable chance of being accepted while in latter iterations, the criterion is increasingly less tolerant in accepting impairments. Usually, the acceptance criterion is the *Metropolis function* which will be presented in due course.

    Based on this criterion's decision, the current solution is either replaced with the new one (i.e., $\psi \leftarrow \psi'$) or not (i.e., ψ is left unchanged).

  - The acceptance criterion is to be modified over the course of iterations. SA is an analogue to the natural crystallization process while cooling. SA's acceptance therefore involves a "temperature" T which isgradually lowered. The effect of this will be discussed in due course.

These steps of suggesting a new candidate solution and deciding whether to accept it for a new candidate solution or not (plus modifying the acceptance criterion), are repeated until some halting criterion is met. For the following implementation, the number of iterations is determined beforehand.

Listing 2.6 provides a pseudocode for the algorithm as presented. As the counter for the iterations starts with the value 2, the number of candidate solutions pro-

```
Initialize ψ with random values;

FOR i := 2 TO I do
    ψ' := ψ;
    j := RandomInteger ∈ [1,...,narg(ψ)];
    z̃ᵢ := RandomValue ∈ [−1,+1];
    ψ'ⱼ := ψⱼ + uᵢ · z̃ᵢ;
    Δℒ := ℒ(ψ') − ℒ(ψ);
    IF Δℒ > 0 THEN
        ψ := ψ'
    ELSE
        with probability p = p(Δℒ,Tᵢ) = exp(Δℒ/Tᵢ) DO
            ψ := ψ'
        END;
    END;

    % New overall best solution?
    IF ℒ(ψ) > ℒ(ψ*) THEN
        ψ* := ψ;
    END;

    Lower temperature: Tᵢ₊₁ := Tᵢ · γ_T;
    If applicable:
        Adjust neighborhood range uᵢ₊₁ := uᵢ · γ_u;
END;
Report best solution ψ*;
```

Listing 2.6: Pseudo-code for GARCH parameter estimation with Simulated Annealing

duced by the algorithm (including the initialization) is equal to I; note also that the iteration loop will be entered only if $I \geq 2$ and skipped otherwise.

A salient ingredient for an efficiently implement HO algorithm are proper values for the algorithm's parameters. For Simulated Annealing, the relevant parameters and aspects include the admitted run time (i.e., the number of iterations), a concept of "neighborhood" (i.e., the modification of ψ), and the acceptance criterion (i.e., a suitable cooling plan). What aspects should be considered in finding values for the respective parameters, will be discussed in the following section.

Unfortunately, there is no unique recipe for how to approach this task. Actually, it can be considered a demanding optimization problem in itself – which is particularly tricky: A certain parameter setting will not produce a unique, deterministic result but rather various results that are (more or less) randomly distributed; the task is therefore to find a combination where the distribution of the reported results is favorable. And as with many demanding problems, there are many possible parameter settings that appear to work equally well, yet it is hard to tell which one is actually the best among them.

Generally speaking, a good parameter setting is one where the algorithm finds reliable solutions within reasonable time. A common way for tuning the algorithm's parameters is to predefine several plausible parameter settings and to perform a series of independent experiments with each of these settings. The results can then be evaluated statistically, e.g., by finding the median or the quantiles of the reported solutions, and eventually select the parameter setting for which the considered statistics are the best; this approach will be used in the following section. Alternative approaches include *response surface analysis* and *regression analysis* where a functional relationship between the algorithm's parameters and the quality of the reported solutions is considered.

For complex algorithms where the number of parameters is high and their effects on the algorithm's quality are highly interdependent, a preselection of plausible parameter values is more difficult; in these circumstances, the parameter values can be found either by a Monte Carlo search – or by means of a search heuristic.

2.4.2 Tuning the Heuristic's Parameters

2.4.2.1 Neighborhood Range

General Considerations Simulated Annealing is a typical neighborhood search strategy as it produces new solutions that are close to the current solutions. It does so by slightly modifying one or several of the decision variables, in the above implementation by adding a random term to the current value: $\psi'_j := \psi_j + u \cdot \tilde{z}$. $\tilde{z}$ is typically a normally or equally distributed random number; here it is chosen to be equally distributed within $[-1, +1]$. The parameter u defines what is considered a neighboring solution: the larger u, the larger the "area" surrounding ψ_j within which the new solution will be, and *vice versa*. Here, "small" and "large" steps have to be seen relative to the variable that is to be changed; hence, the proper value for u will also depend on the magnitude of the different ψ_j's.

Large values for u allow fast movements through the solution space – yet also increase the peril that the optimum is simply stepped over and therefore remains unidentified. Smaller step widths, on the other hand, increase the number of steps necessary to trespass a certain distance; if u is rather small, the number of iterations has to be high. Furthermore, u is salient for overcoming local optima: To escape a local optimum, a sequence of (interim) impairments of the objective function has to be accepted; the smaller u, the longer this sequence is. Smaller values for u demand a more tolerant acceptance criterion which might eventually lead to a perfectly random search strategy, not really different from a Monte Carlo search. With a strict acceptance criterion, small values for u will enforce an uphill search and therefore help to find the optimum close to the current position which might be advantageous in an advance stage of the search.

All this indicates that it might be favorable to have large values for u during the first iteration steps and small values during the last. Also, it might be reasonable to allow for different values for each decision variable if there are large differences in the plausible ranges for the values of the different decision variables.

Finding Proper Values Given the data set for which the GARCH model is to be estimated, the optimal value for $\psi_1 = \mu$ can be supposed to be in the range $[-1, +1]$. Also, it is reasonable to assume that the estimated variance should be non-negative and finite at any point of time. For the variables α_0, α_1, and β_1 in equation (2.1b) (represented in the algorithm by ψ_2, ψ_3, and ψ_4), it is plausible to assume that their values are non-negative, but do not exceed 1; hence, their optimal values are expected in the range $[0, +1]$.

As only one of these decision variables is modified per iteration and the number of iterations might be rather small, we will test three alternatives where u_1 will have an initial value of 0.05, 0.025, or 0.01; the actual modification, $u_1 \cdot \tilde{z}$, will then be equally distributed in the range $[-u_1, +u_1]$.

As argued, it might be reasonable to narrow the neighborhood in the course of the search process. We will therefore test four different versions where u is kept either constant; the value of u in the terminal iteration I is u's initial value divided by

10, 100, or 1 000. The value for u shall be lowered gradually in the course of iterations according to $u_{i+1} = u_i \cdot \gamma_u$. This implies that the parameter γ_u is to be determined according to $\gamma_u = \sqrt[I]{u_I/u_1}$. With the chosen values for u_1 and u_I, γ_u can take the values 1, $\sqrt[I]{0.1}$, $\sqrt[I]{0.01}$, and $\sqrt[I]{0.001}$.

2.4.2.2 The Number of Iterations

General Considerations For some heuristic optimization algorithms, there exist proofs that the global optimum will be identified – eventually. In practical applications, concessions have to be made in order to find solutions within reasonable time. This is primarily done by restrictions on the run time or on the number of iterations. For the latter alternative, common solutions include convergence criteria and upper limits on the number of iterations. Convergence criteria assume that the algorithm has found a solution which is either the global solution – or some local optimum which is unlikely to be escaped and computation time therefore ought to be used for new runs.

As indicated above, selecting the number of iterations is related to finding the parameter for the step size, u, and *vice versa*: The neighborhood range should be large enough that the optimum can actually be reached within the chosen number of iterations and from any arbitrary starting point. Also, for some problems it might be reasonable to have more runs with independent initial guesses for the decision variables, whereas for others it might be advantageous to have fewer runs, yet with more iterations per run.

Finding Proper Values The algorithm will report a solution which it has actually guessed and reached by chance at one stage of the search process. The algorithm, however, does not have a mechanism that guides the search, e.g., by using gradients, estimating the step width with some interpolation procedure or based on past experience. At the same time, we demand a high precision for the parameters. The algorithm is therefore conceded 50 000 guesses before it reports a solution; these guesses can be spent on few independent runs with many iterations or the other way round. We distinguish four versions where all guesses are used on one search run (i.e., the number of iterations is set to $I = 50\,000$), 5 and 50 independent runs with $I = 10\,000$ and $1\,000$ iterations, respectively, and version where no iterative search is performed but all the guesses are used on perfectly random values; this last version corresponds to a Monte Carlo search and can serve as a benchmark on whether the iterative search by SA has a favorable effect on the search process or whether a perfectly random search strategy might be enough.

2.4.2.3 Acceptance criterion

General Considerations In Simulated Annealing the acceptance probability, p, is often determined via the *Metropolis function*, $p = \min\left\{\exp\left(\Delta\mathcal{L}/T_i\right), 100\%\right\}$.[30] For maximization problems, a positive sign for the change in the objective function, $\Delta\mathcal{L} > 0$, indicates an improvement, $\Delta\mathcal{L} < 0$ indicates an impairment. T_i is the analogue for the temperature in iteration i and serves as an adjustment parameter to make the criterion more or less tolerant to impairments: High temperatures push the argument of the $\exp(\cdot)$ expression towards zero and hence the acceptance probability towards 100%, and changes with $\Delta\mathcal{L} \ll 0$ might still be accepted; low temperatures have the adverse effect and make even small impairments unlikely. Improvements, however, are accepted whatever the temperature: as $\exp(\cdot) > 1$ (and therefore exceeds the min-function's limit of 100%) when the argument is positive, the Metropolis function will return an acceptance probability of 100% whenever $\Delta\mathcal{L} > 0$.

Finding good values for the temperature is strongly dependent on what are "typical" impairments. This can be achieved by performing a series of modifications, evaluating the resulting changes in the objective function, $\Delta\mathcal{L}$, and determining the quantiles of the distribution of negative $\Delta\mathcal{L}$'s.

Solving the Metropolis function for the temperature yields $T_i = \Delta\mathcal{L}/\ln(p)$. Hence, during the early iterations, T_i should have values that allow most of the impairments to be accepted with reasonable probability; T_i should therefore be chosen such that a relatively large impairment is accepted with high probability. In the last iterations, only few of the impairments ought to be accepted; here, T_i should have a value such that even a relatively small impairment is accepted with low probability. Once the temperatures for the first and last iterations, T_1 and T_I, have been

---

[30] See also the discussion in section 6.3.1.

u	95%	90%	75%	50%	25%	10%	5%
0.05	−17.70569	*−13.42215*	−8.13720	−3.59098	−0.78051	**−0.21308**	−0.08924
0.025	−17.60417	*−13.09995*	−7.92351	−3.46938	−0.52063	**−0.13813**	−0.05751
0.01	−17.60372	*−13.15639*	−8.06564	−3.58483	−0.35556	**−0.06503**	−0.02704
0.005	−16.94350	−13.06744	−8.07430	−3.41242	−0.19775	**−0.02877**	−0.01147
0.0025	−17.65643	−13.24645	−8.00821	−3.44650	−0.11675	**−0.01526**	−0.00588
0.001	−17.59644	−13.15123	−8.05048	−3.45760	−0.12361	**−0.00617**	−0.00245
0.0005	−17.77629	−13.09759	−7.96761	−3.50134	−0.22632	**−0.00315**	−0.00127
0.00025	−17.63221	−13.09240	−7.93378	−3.41761	−0.17682	**−0.00161**	−0.00066
0.0001	−17.65130	−13.12136	−7.97900	−3.51114	−0.10407	**−0.00063**	−0.00025
0.00005	−17.05599	−12.85940	−7.7959	−3.43883	−0.1256	**−0.00029**	−0.00011
0.000025	−17.49037	−12.95163	−7.91984	−3.45329	−0.11150	**−0.00016**	−0.00006
0.00001	−18.15359	−13.32715	−8.06117	−3.57329	−0.23852	**−0.00007**	−0.00003

Tab. 2.1: Quantiles for modifications with negative $\Delta\mathcal{L}$ for different values of u, based on 10 000 replications each (italics and boldface as explained in the text)

found, the *cooling parameter* $\gamma_T = \sqrt[I]{T_I/T_1}$ can be determined where I is the chosen number of iterations per run. In each iteration, the temperature is then lowered according to $T_{i+1} = T_i \cdot \gamma_T$.

Finding Proper Values   In order to find suitable values for the temperature, the distribution of the potential impairments due to one local neighborhood search step has to be found. This can be done by a Monte Carlo approach where first a number of candidate solutions (that might occur in the search process) are randomly generated and for which the effect of a modification is evaluated. As this distribution depends on what is considered a local neighborhood, Table 2.1 summarizes the quantiles for impairments for the different values of u in the first and the last iteration.

As stated above, the initial values for u were selected from the alternatives [0.05, 0.025, 0.01]. For these three alternatives, the 90% quantiles of the impairments were approximately −13 (see figures in italics in Table 2.1). Hence, if in the beginning, 90% of all impairments should be accepted with a probability of at least $p = 0.5$, then temperature should be set to $T_1 = {-13}/{\ln(0.5)} \approx 20$.

The 10% quantiles of the impairments depend strongly on the value of u; they can be approximated by $-6 \cdot u$ (see figures in boldface in Table 2.1). Hence, if only the

10% of impairments that are smaller than this critical value shall be accepted with a probability of more than $p = 0.1$ in the last iteration, I, the temperature should be set to $T_I = -6 \cdot u_I / \ln(0.1) \approx 2.6 \cdot u_I$. As this shall be the case in the last iteration, the cooling factor is set to $\gamma_T = \sqrt[I]{u_I \cdot 2.6/20}$ where I is the number of iterations.[31]

2.4.3 Results

Based on the above considerations, there are three candidate values for the initial value of u ($u_1 = 0.05, 0.025$ or 0.01) and four alternatives for the terminal value of u ($u_I/u_I = 1/1, 1/10, 1/100$, and $1/1000$; the values for γ_u follow directly according to $\gamma_u = \sqrt[I]{u_I/u_I}$). In addition, we test four different alternatives to use the conceded 50 000 guesses (ranging from a single run with I = 50 000 iterations to 50 000 independent runs with I = 1, i.e., without subsequent iterations[32]), there are 48 different parameter settings to be tested. With each of these combinations, the algorithm was applied to the optimization problem several hundred times, and from each of these experiments, the best of the 50 000 candidate solutions was reported and the distributions of the reported solutions are evaluated. Implemented in Delphi (version 7), the CPU time per experiment (i.e., per 50 000 candidate solutions) was approximately 15 seconds on a Centrino Pentium M 1.4 GHz. Table 2.2 summarizes the medians and 10% quantiles for the deviations between reported solutions and the optimum.

When an adequate parameter setting has been chosen, the algorithm is able to find good solutions with high probability: when the number of iterations is sufficiently high (e.g., I = 50 000) and the parameters for the neighborhood search are well chosen (e.g., $u_1 = 0.025$ and $u_I/u_1 = 0.001$), then half of the reported solutions will have a $\mathscr{L}$ which is at most 0.00001 below the optimum. The best 10% of the solutions generated with this parameter setting will deviate by just 0.000001 or less.

---

[31] A more sophisticated consideration could take into account that during the last iteration, the algorithm has converged towards the optimum and that a random step in this region might have a different effect on $\Delta\mathscr{L}$ as the same modification would cause in a region far from the optimum. Also, the values for p where chosen based on experience from implementations for similar problems; more advanced considerations could be performed. However, for our purpose (and for many practical applications), the selection process as presented seems to generate good enough results.

[32] Note that in listing 2.6, the loop of iterations is not entered when I = 1: To assure that the initial values, too, count towards to the total number of guesses, the loop is performed only I − 1 times.

	u_1	u_τ/u_1	runs × number of guesses per run (I)			
			$1 \times 50\,000$	$5 \times 10\,000$	$50 \times 1\,000$	$50\,000 \times 1$, MC
median	0.05	1	−0.011240	−0.016400	−0.347450	
		0.1	−0.000795	−0.000985	−0.233348	
		0.01	−0.000055	−0.000062	−0.425277	
		0.001	−0.000007	−0.000078	−2.224412	
	0.025	1	−0.004913	−0.005810	−0.489114	
		0.1	−0.000328	−0.000359	−0.650558	−17.895114
		0.01	−0.000024	−0.000103	−5.468503	
		0.001	−0.000008	−1.144997	−23.693378	
	0.01	1	−0.001681	−0.001865	−1.288883	
		0.1	−0.000114	−0.000572	−16.841008	
		0.01	−0.000012	−6.891969	−65.365018	
		0.001	−7.289898	−32.970728	−99.662400	
10% quantile	0.05	1	−0.003348	−0.005210	−0.067568	
		0.1	−0.000309	−0.000363	−0.010878	
		0.01	−0.000020	−0.000025	−0.011645	
		0.001	−0.000002	−0.000008	−0.054592	
	0.025	1	−0.001790	−0.002157	−0.047682	
		0.1	−0.000123	−0.000127	−0.020024	−8.625649
		0.01	−0.000008	−0.000024	−0.090048	
		0.001	−0.000001	−0.008738	−2.413633	
	0.01	1	−0.000614	−0.000749	−0.067869	
		0.1	−0.000041	−0.000112	−0.611584	
		0.01	−0.000004	−0.022991	−15.653601	
		0.001	−0.193161	−2.774787	−37.478838	

Tab. 2.2: 10% quantiles and medians of differences between reported solutions and optimum solution for different parameter settings from several hundred independent experiments (typically 400; MC: 7 800) with 50 000 candidate solutions each

If, on the other hand, the heuristic search part is abandoned and all of the allowed 50 000 guesses are used on generating independent random (initial) values for the vector of decision variables, then the algorithm performs a sheer Monte Carlo (MC) search where there is no neighborhood search (and, hence, the values for u_i are irrelevant) and where, again, only the best of the 50 000 guesses per experiment is reported. The results are by magnitude worse than for SA with a suitable set of parameters (see last column, labeled MC). This also supports that the use of the search heuristic leads to significantly better results – provided an appropriate parameter setting has been selected.

A closer look at the results also underlines the importance of suitable parameters and that inappropriate parameters might turn the algorithm's advantages into their exact opposite. When there are only few iterations and the neighborhood is chosen too small (i.e., small initial value for u which is further lowered rapidly), then the step size is too small to get anywhere near the optimum within the conceded number of search steps. As a consequence, the algorithm virtually freezes at (or near) the initial solution.

However, it also becomes apparent that for most of the tested parameter combinations, the algorithm performs well and that preliminary considerations might help to quickly tune a heuristic optimization algorithm such that it produces good results with high reliability. Traditional methods are highly dependent on the initial values which might lead the subsequent deterministic search to the ever same local optimum. According to Brooks, Burke, and Persand (2001) the lack of sophisticated initializations is one of the reasons why the tested software packages found solutions for the considered problem that sometimes differ considerably from the benchmark. Table 2.3 reproduces their parameter estimates from different software packages[33] together with the benchmark values and the optimum as found by the SA algorithm

---

[33] Brooks, Burke, and Persand (2001) report only three significant figures for the estimates from the different software packages, also the packages might use alternative initializations for σ_0^2. (Our implementation uses the popular approach $\sigma_0^2 = e_0^2 = \frac{1}{T}\sum_{t=1}^{T} e_t^2$ with e_t coming from equation (2.1a).) Reliable calculations of the respective values for $\mathscr{L}$ that would allow for statistically sound tests on the estimation errors are not possible.

Method	$\psi_0 = \mu$	$\psi_1 = \alpha_0$	$\psi_2 = \alpha_1$	$\psi_3 = \beta_1$
Benchmark	−0.00619041	0.0107613	0.153134	0.805974
Heuristic optimization	−0.00619034	0.0107614	0.153134	0.805973
E-Views	−0.00540	0.0096	0.143	0.821
Gauss-Fanpac	−0.00600	0.0110	0.153	0.806
Limdep	−0.00619	0.0108	0.153	0.806
Matlab	−0.00619	0.0108	0.153	0.806
Microfit	−0.00621	0.0108	0.153	0.806
SAS	−0.00619	0.0108	0.153	0.806
Shazam	−0.00613	0.0107	0.154	0.806
Rats	−0.00625	0.0108	0.153	0.806
TSP	−0.00619	0.0108	0.153	0.806

(Software packages: E-Views, Gauss-Fanpac, Limdep, Matlab, Microfit, SAS, Shazam, Rats, TSP)

Tab. 2.3: Results for the GARCH estimation based on the benchmark provided in Bollerslev and Ghysels (1996), the results from the software packages (with default settings) as reported in Brooks, Burke, and Persand (2001)

– which, by concept, uses perfectly random initial values.[34] Unlike with traditional deterministic optimization techniques, this reliability can arbitrarily be increased by increasing the runtime (which is the basic conclusion from convergence proofs for HO algorithms).

2.5   Conclusion

In this chapter, some basic concepts of optimization in general and heuristic optimization methods in particular were introduced. The heuristics presented in this chapter differ significantly in various aspects: the varieties range from repeatedly modifying one candidate solution per iteration to whole populations of search agents each of them representing one candidate solution; from neighborhood search strategies to global search methods, etc. As diverse these methods are, as diverse are

---

[34] In practice, HO techniques do not always benefit when the initial values come from some "sophisticated guess" or another optimization as this often means that the optimizer first and prime task is to overcome a local optimum. On the contrary, heuristically determined solutions might sometimes be used as initial values for traditional methods. Likewise, it might be reasonable to have the fine-tuning of the heuristic's last iterations done by a strict up-hill search.

also their advantages and disadvantages: Simulated Annealing and Threshold Accepting are relatively easy to implement and are good general purpose methods, yet they tend to have problems when the search space is excessively large and has many local optima. Other methods such as Genetic Algorithms or Memetic Algorithms, on the other hand, are more complex and their implementation demands some experience with heuristic optimization, yet they can deal with more complicated and highly demanding optimization problems. Hence, there is not one best heuristic that would be superior to all other methods. It is rather a "different courses, different horses" situation where criteria such as the type of optimization problem, restrictions on computational time, experience with implementing different HO algorithms, the programming environment, the availability of toolboxes, and so on that influence the decision which heuristic to choose – or eventually lead to new or hybrid methods.

The following chapters of this contribution make use of heuristic optimization techniques for approaching problems, merely from the area portfolio management, that cannot be answered with traditional models. The diversity of the problems leads to the application of different methods as well as the introduction of a new hybrid approach. Though the main focus of these applications shall be on the financial implications that can be drawn from the results, there will also be some comparisons of these methods together with suggestions for enhancements.

Chapter 3

Transaction Costs and Integer Constraints

3.1 Introduction

As presented in section 1.1.2, a central simplification in the classical models of Modern Portfolio Theory is that investors face no transaction costs. Although there exit attempts to include proportional[1] or fixed[2] transaction costs, they are usually ignored in the optimization process and explicitly considered only in association with portfolio revision, if at all. When fixed transaction costs or minimum costs are considered, the investor's endowment might become another crucial aspect as the expected return can be influenced significantly by the relative magnitude of these payments.

Based on Maringer (2002a), where, to our knowledge, the aspects of different transaction costs, non-negativity constraints and integer constraints are jointly investigated for the first time, this chapter will primarily investigate the relationship between transaction costs, type of costs, and initial endowment of the investor on the one hand and the optimal portfolio structure on the other. In section 3.2 the optimization problem will be formalized and the solution approach and the used data will be presented. Based on an empirical study for a DAX investor, section 3.3 presents the central results and section 3.4 discusses some of the implications for portfolio management. Section 3.5 concludes.

---

[1]   See, e.g., Pogue (1970).

[2]   See, e.g., Brennan (1975).

3.2   The Problem

3.2.1   The Optimization Model

Typically, transaction costs are proportional to the traded volume and are therefore variable. They may have a lower limit (i.e., minimum transaction costs), and they may also come together with fixed costs, e.g., fees per order. Fixed costs only are not very common, yet some brokers do offer cost schemes where the fees are fixed within certain (more or less broad) ranges of the traded volume.

Let $n_i \in \mathbb{N}_0^+$ be the natural, non-negative number of asset $i \in [1,...,N]$ and S_{0i} its current price. If the investor faces proportional costs of c_p and/or fixed minimum costs of C_f, investing into asset i comes with transaction costs C_i of

$$C_i = \begin{cases} C_f & \text{fixed costs only} \\ c_p \cdot n_i \cdot S_{0i} & \text{proportional costs only} \\ \max\{C_f, c_p \cdot n_i \cdot S_{0i}\} & \text{proportional costs with lower limit} \\ C_f + c_p \cdot n_i \cdot S_{0i} & \text{proportional plus fixed costs} \end{cases} \qquad (3.1)$$

The total payments associated with the purchase of n_i stocks i are then $n_i \cdot S_{0i} + C_i$. The investor has an initial endowment of V_0 that shall be invested into stocks. Due to the indivisibility of stocks via the integer constraint on n_i, the actual amount spent on stock purchases will not equal V_0 exactly, and the remainder

$$R_0 = V_0 - \sum_{i=1}^{N} (n_i \cdot S_{0i} + C_i)$$

can be invested at the settlement account without cost and at a safe rate of r_s. It is assumed that the safe asset also has a non-negativity constraint, i.e., $R_0 \geq 0$, debt-financed stock purchases are therefore excluded.

If there are no transaction costs at the end of the period, the effective return on the initial endowment for the period will be

$$r_p^{eff} = \frac{\sum_{i=1}^{N} (n_i \cdot S_{0i} \cdot (1+r_i)) + R_0 \cdot (1+r_s)}{V_0} - 1$$

where r_i is the expected rate of return of asset i. Let

$$x_i = \frac{n_i \cdot S_{0i}}{V_0} \iff \sum_{i=1}^{N} x_i = 1 - \frac{R_0 + \sum_{i=1}^{N} C_i}{V_0},$$

then the expected risk premium and volatility, respectively, for the portfolio after transaction costs can be written as

$$r_P^{eff} - r_s = \sum_{i \in P} \left(x_i \cdot (1 + r_i) - \frac{C_i}{V_0} \cdot (1 + r_s) \right)$$

$$\sigma_P = \sqrt{\sum_{i \in P} \sum_{j \in P} x_i \cdot x_j \cdot \sigma_{ij}}$$

where σ_{ij} is the covariance of i's and j's expected return.

As argued in section 1.2.1, an investor will want to maximize the ratio between (expected) risk premium and total risk associated with the investment, also known as *Reward to Variability Ratio* or *Sharpe Ratio* (SR). Under transaction costs, this ratio has to be redefined as

$$SR_P^{eff} = \frac{r_P^{eff} - r_s}{\sigma_P}.$$

The optimization problem for a myopic investor can therefore be summarized as follows:

$$\max_{n_i} SR_P^{eff} = \frac{r_P^{eff} - r_s}{\sigma_P}$$

subject to

$$r_P^{eff} - r_s = \sum_{i \in P} \left(x_i \cdot (1 + r_i) - \frac{C_i}{V_0} \cdot (1 + r_s) \right)$$

$$\sigma_P = \sqrt{\sum_{i=1}^{N} \sum_{j=1}^{N} x_i \cdot x_j \cdot \sigma_{ij}}$$

$$x_i = \frac{n_i \cdot S_{0i}}{V_0}$$

$$n_i \in \mathbb{N}_0^+$$

$$R_0 = V_0 - \sum_{i=1}^{N} (n_i \cdot S_{0i} + C_i) \geq 0$$

$$C_i = \begin{cases} C_i\left(n_i, S_{0i}, C_f, c_p\right) & n_i > 0 \\ 0 & \text{otherwise} \end{cases}.$$

The additional objective of keeping as little cash in the settlement account as possible is not stated explicitly (e.g., by adding a *punishment term* to the objective function), but will be incorporated in the optimization process.

3.2.2 The Heuristic

The portfolio selection problem presented above will be approached with an adapted version of *Simulated Annealing*[3] (*SA*), a method successfully applied to portfolio optimization in Chang, Meade, Beasley, and Sharaiha (2000) or Crama and Schyns (2003) and a method closely related to Threshold Accepting, the method used for portfolio selection in Dueck and Winker (1992).

In this application, the algorithm starts with an arbitrary portfolio that is valid with respect to the constraints. According to the principle of SA, in each iteration step slight random changes of the current portfolio structure are suggested by first selecting one asset i and lowering its weight by a random amount and adding the "saved" amount (including avoided transaction costs) to the remainder R_0. Next, another asset j is selected and n_j increased by as much as possible given the current amount R_0. If this new solution has a higher SR_P^{eff} than the previous one, then this new solution is accepted. If the value of the objective function has decreased, then the suggested restructuring is rejected stochastically; the sharper the decline in SR_P^{eff} and the more progressed the algorithm is (i.e., the lower the current temperature T_t), the higher the chances that the suggested changes are not accepted and the previous portfolio structure is kept. Listings 3.1 and 3.2 summarize the pseudo-code. The algorithm is linear in the number of iterations, I, and quadratic in the number of different assets, N, due to the computation of a candidate portfolio's variance; hence, the computational complexity of the algorithm is $\mathcal{O}\left(\mathrm{I} \cdot N^2\right)$.

According to the acceptance criterion of SA, changes are kept rather generously during the first iterations, implying that, although improvement are always more

---

[3] See Kirkpatrick, Gelatt, and Vecchi (1983) and the presentation in section 2.3.1.

```
n := RandomInitalStructure;
initialize parameters
    T₁ (Temperature), γ (CoolingFactor), neighborhood range;

FOR i = 1 TO MaxIterations (I)
    perform neighborhood search (→ Listing 3.2);
```
$\Delta SR^{eff} := SR^{eff}_{n'} - SR^{eff}_{n} ;$

$ReplacementProbability := \min\left(1.\exp\left(\frac{\Delta SR^{eff}}{T_i}\right)\right);$

```
    with probability ReplacementProbability DO
        n := n';
        check whether new elitist (overall best solution) has been found:
```
$\qquad\qquad$IF $SR^{eff}_n > SR^{eff}_{n^*}$ THEN

$\qquad\qquad\qquad n^* := n ;$

```
            END;
    END;
```
$\qquad T_{i+1} := T_i \cdot \gamma;$
```
END;

report n*;
```

Listing 3.1: Pseudo-code for the main Simulated Annealing routine

```
randomly select assets i (where nᵢ > 0) and j (where j ≠ i);
```
$\Delta n_i := \text{round((rand*maxChange)} * x_i * V_0 / S_{0i});$

$n'_i := n_i - \Delta n_i ;$

```
compute R'₀ (including saved transaction costs);
determine maximum Δnⱼ without violating R₀ ≥ 0,
```
$n'_j := n_j + \Delta n_j ;$

```
return n;
```

Listing 3.2: Pseudo-code for the local search routine

likely to be accepted than impairments, the change of the portfolio structure has a considerable degree of randomness which fosters the search for the core structure without getting stuck in a local optimum. During later iterations, however, the temperature T_i has been lowered substantially, and the algorithm turns more and more towards a hill-climbing search strategy where changes for the better are definitely preferred over reductions in the objective function. At this stage of the algorithm, the initially found core structure of the portfolio is refined. At any stage, there is a certain chance to move away from an already found good solution, the algorithm therefore records the best known solution found over all passed iterations, n^*, which is ultimately reported. In order to reduce the risk of reporting a local optimum,[4] each of the optimization problems has been solved independently several hundred times, in addition there was a series of concluding runs were the initial solution was set to the overall best known solution from previous runs for this problem and "closely related" problems, i.e., with close values for V_0, c_p, and C_f. The following evaluations are based on the best overall result for each problem setting which can therefore be assumed to either represent the global optimum or to be at least sufficiently close to.

3.2.3   The Data

The empirical study in this section is based on the 30 stocks represented in the German stock index DAX. Using daily returns for the period Dec. 1, 1998 – Dec. 29, 2000, the covariances and volatilities, σ_{ij} and σ_i, respectively, and the historic correlation coefficients between the DAX and the stock returns, ρ_{iDAX}, were computed. The expected covariances (volatilities) for the optimization period are the annualized historic covariances (volatilities), and the expected returns are estimated with a standard CAPM approach.[5] To reflect the market situation for the chosen time frame (yet without loss of generality), the safe return was assumed to be $r_s = 4.5\%$ and the DAX's expected risk premium was set to $(r_{DAX} - r_s) = 6\%$. The estimated returns are therefore generated by

$$r_i = 0.045 + 0.06 \cdot \beta_i \text{ with } \beta_i = \frac{\sigma_i \cdot \rho_{iDAX}}{\sigma_{DAX}}$$

---

[4] See also section 2.3.1.

[5] See section 1.2.2.

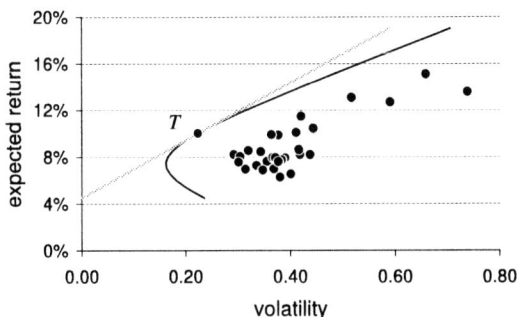

Fig. 3.1: Risk and expected return of considered stocks and efficient lines with (gray) and without (black) safe asset

where σ_{DAX} is the expected volatility of the DAX, based on its historic value. Figure 3.1 depicts these expected returns and volatilities. Given the actual market developments it shall be emphasized that for the empirical study, these figures are estimated based on information available to investors at the end of the year 2000 and that they are considered to be exogenous to the optimization process: the optimization itself concentrates on the portfolio selection problem but not on the prediction. The turbulences on the international and the German stock markets would lead, of course, to readjusted estimates for risks, returns, and hence Sharpe Ratios; subsequent tests with modified estimates, however, did not affect the qualitative results of the following analyses. To underline the generality of these results, the stocks will be made anonymous in the following evaluations.

For the computational study, the following values were considered: The initial endowment, V_0, ranged from € 500 up to € 10 000 000; the proportional costs ranged from 0 up to 5%, the fixed and minimum costs, respectively, ranged from € 0 to € 100. The upper limits for the costs had been chosen rather high for stocks, yet are not unusual for other investments such as international funds. In addition, these broad bandwidths allow a clearer look at the underlying effects and mechanics between costs and optimal investment decisions.

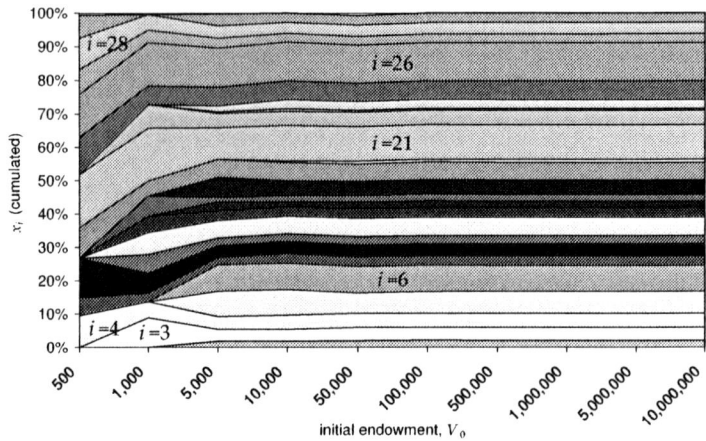

Fig. 3.2: Optimal portfolio structure for an investor with initial endowment V_0 without fixed or proportional transaction costs

3.3 Results from the Empirical Study

3.3.1 The Influence of Non-Negativity and Integer Constraints

On Dec. 29, 2000, the closing quotes of the 30 DAX stocks ranged from € 16.50 up to € 400.70 with an average of € 79.24. To an investor with an initial endowment of just $V_0 = $ € 1 000, the cheapest stock represents 1.65% of her endowment, the most expensive one almost half of the amount she is willing to invest. This is also the bandwidth of the granularity up to which the investor can split her endowment. Hence, for small investors, the usual assumption of arbitrarily divisible assets is not a very realistic one. Large investors, on the other hand, come much closer to this assumption as their step sizes are smaller by the magnitude their initial endowment is larger. Hence, one might expect that integer constraints on the number n_i held of asset i in the portfolio will affect small investors significantly more than large investors. Figure 3.2 depicts the optimal portfolio structure depending on the initial endowment when there are neither fixed nor proportional transaction costs.

The results indicate that, for the given market, the integer constraint has no noticeable effect on the optimal solution only when the amount invested is € 50 000 or more. Although there are slight alterations for higher V_0, the differences in the respective values for SR_P^{eff}, however, are negligible, and the solutions correspond more or less to the optimal result without integer constraint. Here, it is the non-negativity constraint that is the reason for not including all available assets into the portfolio:[6] under the assumed risk and return structure, ideally five of the assets should be sold short – which, however, is prohibited by assumption. Though it influences the portfolio structure, the non-negativity constraint's harm on the Sharpe Ratio is very limited: When abandoning this constraint, an investor with an endowment of $V_0 =$ € 100 000 can increase her SR_P^{eff} from 0.24612 to 0.24625, i.e., by just about 0.05%.[7]

For an investor with $V_0 =$ € 10 000 or less, the effects of the integer constraint become apparent. Stock with high prices (in particular the one with $S_{0i} =$ € 400.70) are likely to be eliminated as the investor cannot buy the (small) quantity that would contribute to the optimal solution the best and are therefore replaced with other assets. As a consequence, the number of different assets, included in the portfolio, decreases. Whereas large investors should pick 25 different stocks, those with low V_0 would be best off with as little as ten to fifteen different stocks.

It is noteworthy, however, that the changes in the portfolio structure due to the lower V_0 are not always derivable from the theoretical solution for frictionless trading. Whereas the weight of asset $i = 28$ increases with decreasing V_0, the weights of assets $i = 21$ and $i = 26$ remain virtually constant. Since all three of these assets are quoted at more or less the same price, these effects on the changing weights are caused by the return and risk structure in relationship to the remaining other assets. This implies that the simple rule, often applied in practice, of first finding the optimum without the integer constraint and then determining the actual quantities n_i by rounding, can lead to inferior solutions.[8] A numerical example underlines this effect: When including the integer constraint, for an investor with $V_0 =$ € 5 000, the optimal portfolio will have $SR_P^{eff} = 0.24589$, for an investor with $V_0 =$ € 500, it is $SR_P^{eff} = 0.23466$. If, however, the second investor would have determined her asset weights by simply "downsizing" the first investors portfolio and rounding the weights, the

---

[6] According to Best and Grauer (1992), following Best and Grauer (1985) and Green (1986) and the literature given therein, it is unlikely to have optimized portfolios where all weights are positive when generated from historical data. See also Best and Grauer (1991).

[7] The magnitude of this effect depends on the market structure as will be seen in chapter 4.

[8] See, e.g., chapter 13 in Hillier and Lieberman (2003).

result would have been $SR_{\mathcal{P}}^{eff} = 0.21975$. In other words, the disadvantages of small initial endowment on the limited opportunities of diversification can be avoided to some extent when considered within the optimization process and not afterwards.

If there are no transaction costs, investors with an initial endowment of $V_0 =$ € 5 000 or less will be affected noticeable by the integer constraint, and they should rather account for this fact when determining the actual asset weights. Investors with an endowment of up to $V_0 =$ € 50 000 will also be effected by the integer constraint, yet with rather limited consequences. Only for investors with higher V_0 the effect of including an explicit integer constraint will vanish.

3.3.2   Simple Transaction Costs

3.3.2.1   Fixed Costs Only

When introducing fixed costs of $C_i = C_f \forall i : n_i > 0$ diversification comes with higher transaction costs the more different assets are included in the portfolio. This implies that "aggregation" and spending a certain amount on one asset rather than two similar stocks will cut the resulting costs in half. Small investors can therefore be expected to (even more) reduce the number of different assets in their portfolios as to them fixed costs are relatively higher than to large investors. In other words, the contribution of an asset to the portfolio's Sharpe Ratio has to be large enough so that it is not outweighed by the additional relative costs due to the inclusion of this asset. Both the reduced diversification and the relatively high costs will aggravate and make a small investors' $SR_{\mathcal{P}}^{eff}$ rather sensitive towards fixed costs. For large investors, on the other hand, the fixed amount of C_i per included different asset will be relatively small; the effect on the Sharpe Ratio will therefore be less severe than for small investors.

Figure 3.3 presents the effect of different levels of fixed costs on the optimal portfolio structure for an investor with an initial endowment of $V_0 =$ € 50 000. As can be seen, introducing fixed costs of as little as $C_f =$ € 5 causes the investor to include only half of the different assets she would hold when there were no transaction costs. This reduction affects merely those assets that would have low weights in a cost free portfolio. Increasing C_f leads to further decline in the portfolio's diverseness and to a

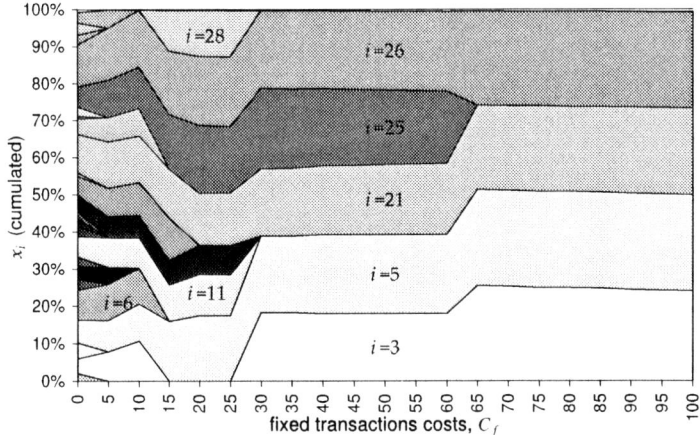

Fig. 3.3: Optimal portfolio weights for an investor with $V_0 = € 50 000$ for different levels of fixed costs without proportional costs

shift in the remaining assets' weights. However, these transitions are not necessarily smooth: Some securities, such as $i = 28$, can even gain substantially in weight while being excluded both under lower and higher costs. The reason for this is their ability to function as a substitute for a group of other assets under certain cost regimes: exchanging a bundle of assets for a single asset with equal properties might be a good idea when the avoided costs more than offset the disadvantages in this assets risk/return structure before costs. Ultimately, assets of that kind will either be excluded as well (as is the case for $i = 28$) – or might even become one of the few "survivors" (as is the case for $i = 3$).

Investors with lower initial endowment face similar effects of reducing the number of different assets, yet already at lower fixed costs. The larger the initial endowment, on the other hand, the less distinct the effects of fixed costs become; nonetheless, even with V_0 as high as $€ 10 000 000$, one or two of the otherwise included assets might fall out of the optimal portfolio. Figure 3.4 illustrates this relationship.

As expected, fixed costs do have a negative effect on the investors' optimal Sharpe Ratios and small investors are affected the most. In some cases, the Sharpe Ratio might even become negative, implying that in these situations the investor would be better off when not holding any stock portfolio but keeping her (low) initial endowment in the bank account as the portfolio's expected return after transaction costs

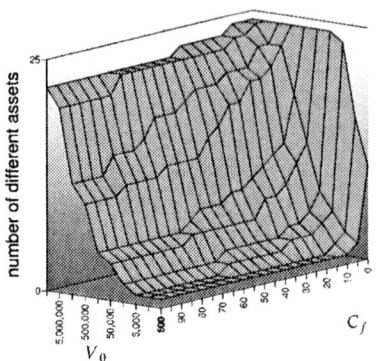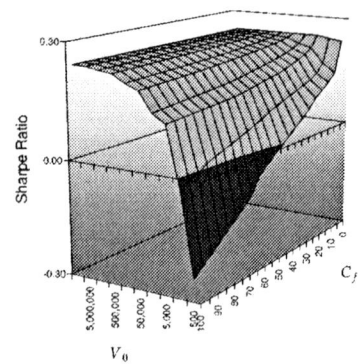

Fig. 3.4: The effects of fixed costs on the number of assets and the Sharpe Ratio

would be lower than the safe return. The higher V_0 the lower the decrease in $SR_\mathcal{P}^{eff}$: large investors buy larger quantities, and the fixed payments for transaction costs have a relatively insignificant effect on these investors Sharpe Ratios after transaction costs. The still noticeable decline in the number of different assets for investors with an endowment of up to $V_0 = 1\,000\,000$ can be offset and does therefore not really show in the respective values for $SR_\mathcal{P}^{eff}$, as can be seen from Figure 3.4

3.3.2.2   *Proportional Costs Only*

Whereas fixed transaction costs can (more or less) be avoided by reducing the number of different assets in the portfolio, proportional costs depend exclusive on the transaction volume. Hence, substituting some of the securities by higher volumes in already included assets will not lead to a reduction in the overall transaction costs – apart from insignificant changes due to integer constraint. Much akin to fixed costs, however, the marginal contribution of an included security to the overall diversification will be reduced and the portfolio's Sharpe Ratio might even become negative. As a consequence, proportional transaction costs, too, can be expected to lead to less diversified portfolios and reduces Sharpe Ratios.

Figure 3.5 illustrates the influence of proportional transaction costs on the optimal portfolio structure for an investor with an initial endowment of $V_0 = €\,50\,000$. As can readily be seen, increasing the proportional cost rate leads to a gradual reduction of included stocks and a shift of the portfolio weights. Whereas increasing

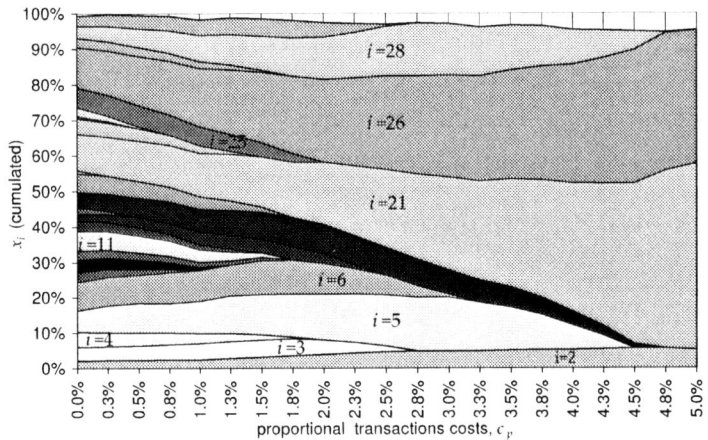

Fig. 3.5: Optimal asset weights for an investor with $V_0 = €\,50\,000$ with proportional transaction costs only

the proportional costs from 0 to 0.5% (1%) will lead to a reduction from originally 24 down to 21 (19) different assets.

Again, the optimal weights under transaction costs are not necessarily obvious from the weights for the unconstrained problem: Whereas assets $i = 4$, 25 and 28 are almost identical in weight in the absence of costs, the weight of the first one decreases continually, the weight of the second one remains almost constant and then vanishes quickly, and the weight of the third one increases noticeable yet eventually decreases and vanishes, too. Compared to the effects of fixed transaction costs only (see Figure 3.3), some of the assets now show contrary behavior. Asset $i = 2$, e.g., has a below average weight when there are no transaction costs, but steadily gains weight when the proportional costs rate rise; under a fixed costs regime, however, this very title would have been excluded already at a rate of $C_f = €\,5$. Asset $i = 3$, on the other hand, is eventually excluded under proportional costs, but turns out to be one of the view securities included under high fixed costs whereas under low fixed costs it might be replaced with asset $i = 28$. The main reason for this opposing effects of different cost regimes is the fact that under proportional costs, the effective rate of return depends on the expected return of the asset and the proportional cost rate, whereas under fixed costs, it also depends on the share the asset takes in the portfolio which in return is influenced by the initial endowment. This means

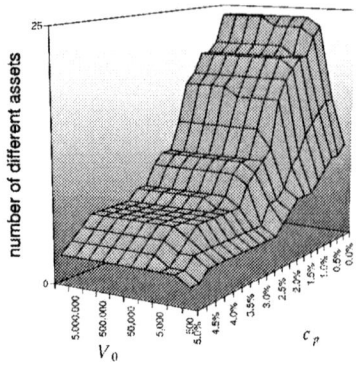 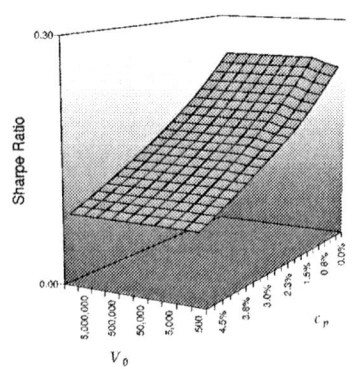

Fig. 3.6: The effects of proportional costs on the number of assets and the Sharpe Ratio

that given proportional costs, the expected return after costs is independent of the weight it has in the portfolio; given fixed costs, however, reducing an asset's weight also means reducing its expected return after costs and *vice versa*. If therefore a single asset turns out to be a potential substitute for a group of other assets all of which have small weights, than eliminating this group and increasing the single asset's weight has a positive effect on this asset's expected effective return. When the costs are proportional to the traded volume, considerations of this kind are irrelevant, and "aggregating" or switching between single assets and bundles is no longer an adequate measure to avoid costs. The changes in the weights are therefore much smoother in a proportional cost regime.

Figure 3.6 shows the decline in the number of different included assets in dependence of V_0 and c_p. When compared to the fixed cost regime, it is remarkable that now the differences between small and large investors have largely vanished and that an investor with $V_0 = 10\,000\,000$ appears to have a portfolio quite similar to an investor with just $V_0 = 10\,000$. This also shows in the similar SR_p^{eff} the two investors will expect (see Figure 3.6, right panel).

In practice, proportional costs of more than 1% are rather unusual for domestic stock purchases, but can be substantially higher for investments into securities on foreign markets or funds where issue surcharges as well as management fees have to be paid. The seemingly generous upper limit of $c_p = 5\%$ appears therefore justified when conclusions for securities of this kind are to be drawn. In this respect, the empirical results from this study would indicate that *ceteris paribus* an individual investing into funds should focus on a rather limited number yet well-chosen combination of different securities.

3.3.2.3 Comparison of Simple Cost Schemes

Comparing the results for fixed costs with those for proportional costs also allows some conclusions on preferences and effects from changing the type of costs. This shall be illustrated with a numerical example for an investor with an initial endowment of $V_0 = 50\,000$.

If there are no transaction costs, this investor will hold 24 different assets. If, however, she faces fixed costs of $C_f = 15$ per different asset, she will hold just eight different securities, i.e., two thirds of the originally attractive securities won't be included in the optimal portfolio. In this case, the overall cost for this portfolio will be $8 \cdot 15 = 120$. In proportion to the amount invested, this corresponds to pro-rata costs of 0.24%. The expected Sharpe Ratio after costs will be $SR_p^{eff} = 0.22960$.

If, on the other hand, the investor had to face proportional costs of 0.25% of the purchased volume yet neither fixed nor minimum transaction costs, than the investor will optimally purchase a portfolio that contains 23 different assets and that has an initial value of € $49\,863.24$; the payments for transaction costs sum up to 124.66, and a remainder of only $R_0 = 12.10$ remains on the settlement account. With (more or less) the same overall costs, the investor is now able to diversify much better than in the fixed costs case, which is apparent from the significantly higher Sharpe Ratio of $SR_p^{eff} = 0.23393$. In this case, a change of the cost structure is therefore advantageous for the investor because she can get higher benefits (in terms of the reward-to-variability ratio) for the same costs. Assuming that the costs are exclusively for the benefit of the financial services provider, things look different from their point of view: the number of different orders has almost tripled, yet the charges remain almost unchanged.

When the purchase of 23 different stocks costs € 124.66 (as is the case for the just mentioned example with $c_p = 0.25\%$), then this corresponds to € 5.42 per different stock. This does not imply, however, that changing from the proportional to a fixed costs system where $C_f = 5.42$ would lead to the same investment decision: In the latter case, the investor would select just eleven different stocks for her portfolio, and for the overall costs of $11 \cdot 5.42 = 59.62$ would get a portfolio that had a

Sharpe Ratio of $SR_p^{eff} = 0.23780$ – which is above the one for $c_p = 0.25\%$. Again, the possibility of reducing costs by aggregation and substituting sub-portfolios with single assets is utilized. If a bank therefore considers changing its charging system from proportional to fixed costs, using the current pro-rata figures for a benchmark in the new system might result in a lasting reduction in the number of orders and an increase in the volume per order; the benchmark for sustaining the current revenues from transactions would therefore be the value for the initial considerations in this numerical example, namely $C_f = 15$ which is almost three times as high as the equivalent fixed pro-rata figure under a proportional costs regime.

As pointed out previously, under fixed costs, the initial endowment has a major effect on the portfolio structure. Investors with different V_0 will therefore react differently on changes in the cost system, the composition of a bank's clients' portfolios and endowments are therefore crucial for a reliable evaluation of the relationship between changes in the costs structures and the revenues. Nonetheless, this rather simple example captures the basic mechanics behind.

3.3.3   Compound Transaction Costs

3.3.3.1   *Proportional Costs with Lower Limit*

If the purchase is associated with variable costs which, however, must not fall below a minimum amount, than including $n_i > 0$ assets i in a portfolio causes payments of

$$C_i = \max\{C_f, c_p \cdot n_i \cdot S_{0i}\} = \max\{C_f, c_p \cdot x_i \cdot V_0\}.$$

Hence, fixed (minimum) costs are to be paid whenever x_i is positive yet below the critical value of

$$C_f = V_0 \cdot x^* \cdot c_p \Rightarrow x^* = \frac{C_f}{V_0 \cdot c_p}.$$

Ceteris paribus, this critical value will be the lower, the lower C_f is; the better endowed the investor is – and the higher the rate of the proportional costs is. As has been argued in the previous section, when there are only fixed costs, small investors will differ substantially in their decision from large investors. Now, one also has to consider that the fixed (minimum) costs have a low chance of coming into effect

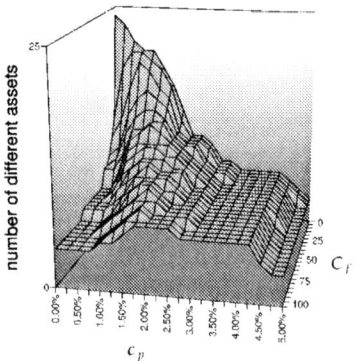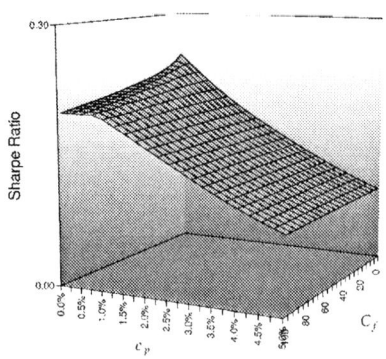

Fig. 3.7: Effects of proportional costs c_p with a lower limit of C_f for the case $V_0 = 50\,000$

and are therefore more unlikely to influence the investor's decision when the proportional costs are high. At the same time, this also means that the "advantage" of fixed costs, namely avoiding costs by focusing on a small number of different assets and aggregating, is lost. Hence, relatively low minimum costs, i.e., low x^*, will lead to an investment behavior similar to the case of variable costs only, whereas a comparatively low rate for the proportional costs will increase x^* and subsequently lead to decisions similar to the case of fixed costs only.

This relationship also explains a phenomenon that, at first sight, might be surprising (Figure 3.7): For given minimum costs, the number of included assets might increase when c_p increases. When the minimum transaction costs are $C_f = €\ 15$, an investor with $V_0 = €\ 50\,000$ will select just eight different stocks for her portfolio when there are no variable costs, yet nine different assets when $c_p = 0.25\%$ and 16 when $c_p = 1\%$, and only when $c_p \geq 2.75\%$ she will hold less than eight different assets. Though the figures are different for investors with different V_0, the effect remains basically the same. Only when the minimum costs are rather high, the introduction of low proportional costs does not affect the investment decision noticeably. Otherwise, the introduction of variable costs eliminates the advantage of aggregating, and the reduction in the expected effective returns can be (largely) offset by a higher degree of diversification. Nonetheless, increasing c_p has always a negative effect on SR_P^{eff} (see Figure 3.7).

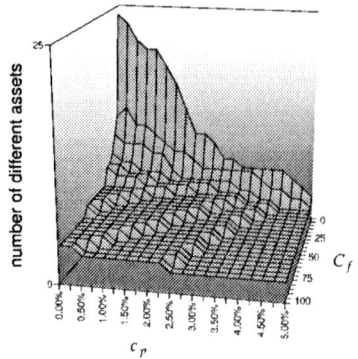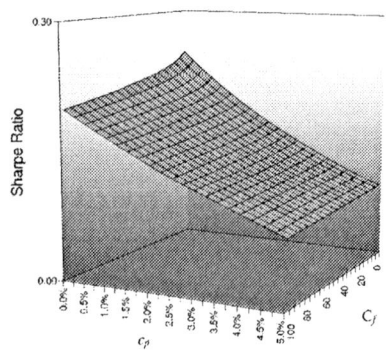

Fig. 3.8: Effects of proportional costs c_p plus fixed charges of C_f for the case $V_0 = 50\,000$

3.3.3.2 Proportional plus Fixed Costs

If a stock purchase causes proportional plus fixed costs, then the effects of the respective type of costs are combined. This means that increasing variable costs (with equal fixed costs) as well as increasing fixed costs (with equal proportional costs) will reduce the opportunities for diversification and the number of included different assets will decline. Figure 3.8 depicts this effect for an investor with $V_0 = €\,50\,000$ – and also gives evidence that the integer constraint might make a difference: for $C_f = €\,5$, increasing the proportional costs, c_p, from 0 to 0.25% leads to an optimal solution where the number of assets included also increases from twelve to thirteen. Unlike in other circumstances, in this particular case the effect of the additional asset is rather academic: the SR^{eff} is just 0.00001 higher than it would be for the optimal combination when the original twelve assets were kept and rearranged given the same cost structure.

Figure 3.8 confirms the expectation that the case where proportional plus fixed costs have to be paid has the most serious effect on the expected SR_P^{eff} of all cost structures since in this case the payments for transaction costs are here the highest in absolute figures.

3.4 Consequences for Portfolio Management

For any of the discussed types of costs, the portfolio structures of different investors should ideally differ not only in the number of included assets but merely in the weights assigned to the different stocks. This implies that inferior solutions can only be avoided when the portfolio selection process does consider the costs the investor faces and her initial endowment. Consequently, individual and tailor-made portfolio management might be reasonable also in a world where all investors have homogeneous expectations about the available assets' risks and returns. The *separation theorem* for a perfect market with several risky assets and one risk-free asset, according to which the risky assets' weights have the same proportions for any investor and their optimization is therefore separated from the individual's investment decision,[9] no longer holds when there are transaction costs.

This shall be illustrated by considering three different investment opportunities:

Individual selection: According to the optimization model presented in the previous sections, the investor optimizes an individual portfolio given the constraints.

Market fund: We assume that there is a readily available investment fund offered that faces neither non-negativity nor integer constraints, the weights are therefore optimized for the respective unconstrained optimization problem without transaction costs. Costs have to be paid only when investing into this fund by considering it to be a single asset causing the same costs as when buying the respective quantity of any other single stock. In addition, we assume frictionless trading, i.e., the investor can buy any quantity that fits her budget constraint the best.

Replication: The investor finds the optimal solution for the unconstrained optimization problem and tries to translate it according to her personal situation by calculating the actual number of stock i, n_i, for each i based on the theoretical weights, the stock prices, initial endowment and costs. The values for n_i are scaled and rounded such that neither the integer nor the budget constraint is violated and as much of the initial endowment as possible is spent. In the sense of comparability, short sales are excluded.

---

[9] See section 1.1.2.3.

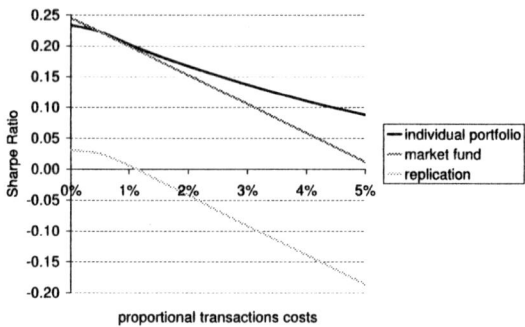

Fig. 3.9: Sharpe Ratio for an investor with $V_0 = €\ 50\,000$ and minimum transaction costs of $€\ 10$

For an investor with, again, $V_0 = €\ 50\,000$ who faces fixed minimum costs of $C_f = €\ 10$, Figure 3.9 depicts the Sharpe Ratio she can expect for different rates of proportional costs. As can easily be seen, simple replication of a portfolio optimized with no constraints leads to severely inferior solutions which are clearly outperformed by both other alternatives. The shortcomings of this solution steam not only from the securities with low weights that would otherwise be excluded because of the minimum costs or with too low effective returns when the proportional costs are considered, but also from the inappropriate weights of the other assets, as will be discussed later. Individual selection can and does account for this fact, this alternative therefore finds definitely better solutions. The investor should therefore abandon the idea of simply replicating a portfolio optimized for a perfect market.

When comparing the individual selection to the market fund, the latter might be the preferred solution when the proportional costs are low (if existent at all) and the investor had therefore to pay the minimum costs several times for all the different assets in her portfolio, but only once when investing into the fund. The region where the fund is the better alternative than the individual selection will be the larger the higher the minimum costs, the lower the initial endowment and the lower the rate

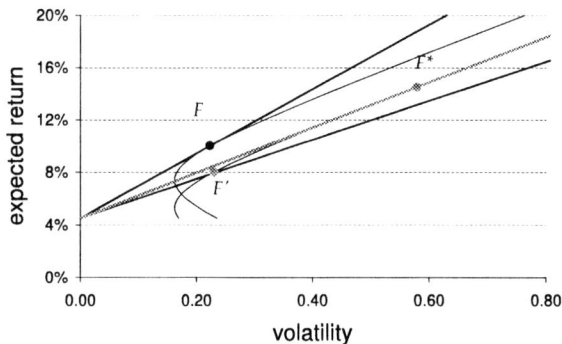

Fig. 3.10: Optimal funds without transaction costs (F) and with proportional costs of $c_p = 0.02$ (F^*)

of proportional costs. The higher c_p, on the other hand, the more advantageous the individual selection will be.[10]

The reason for this becomes apparent from a closer look at the fund's structure and the way it is optimized. For the considered example, the minimum costs will be exceeded by the proportional whenever $c_p > c_f/V_0 = 0.02\%$, and the investor will then be able to invest a net amount of $V_0/1+c_p$ into the fund F. The expected return of this fund after transaction costs, r_F^{eff}, will therefore be

$$V_0 \cdot \left(1 + r_F^{eff}\right) = \frac{V_0}{1+c_p} \cdot (1 + r_F) \Rightarrow r_F^{eff} = \frac{1+r_F}{1+c_p} - 1,$$

where $r_F = \sum_i x_i^F \cdot r_i$ is the expected return of the fund and $x_i^F : \sum_j x_j^F = 1$ are the respective weights of the unconstrained optimization problem, i.e., the weights of the tangency portfolio in the Tobin model[11] for a perfect market. After substituting and rearranging, r_F^{eff} can be rewritten as

$$r_F = \sum_i \left(x_i^F \cdot \frac{r_i}{1+c_p} \right) - \frac{c_p}{1+c_p}.$$

---

[10] See also Sharpe (1991) who considers restrictions on negative holdings as only impediment. He finds that in this case, "the market portfolio may not be efficient" and that "overall welfare may be lower than it would be if the constraints on negative holdings could be reduced or removed" (p. 505).

[11] See the discussion in section 1.1.2.3.

In the Tobin model, the portfolio weights are not linearly homogeneous in the individual securities' returns, hence adjusting all expected returns r_i by $1/(1+c_p)$ would demand re-optimizing and calculating the new weights x_i^{F*} – which, however, is not done when sticking to the original fund F. The higher c_p, the greater the resulting deviations will be. This can also be seen from Figure 3.10 for the DAX investor in this empirical study. When there are no transaction costs, F would be the optimal solution. Introducing proportional costs, however, of $c_p = 2\%$ moves the Markowitz efficient line down, and the original tangency portfolio has now become F'. Since the safe return remains unaffected by the proportional costs, F' is no longer the optimal solution, and investing into F^* would yield a higher effective Sharpe Ratio which, in this diagram, corresponds with the slope of the respective tangency lines. Investing in the fund with the original weights x_i^F leads to an actual effective Sharpe Ratio of

$$SR_{F'}^{eff} = SR_F^{eff} - \frac{1+r_s}{\sigma_F} \cdot \max\left\{\frac{C_f}{V_0}, c_p\right\},$$

where SR_F^{eff} and σ_F are the Sharpe Ratio and volatility, respectively, of the unconstrained tangency portfolio in a perfect market without costs.

The Sharpe Ratio for an investment into the fund therefore linearly declines when $c_p > C_f/V_0$ is increased. Individual portfolio selection under transaction costs can avoid this specification error and find a better portfolio, corresponding to a tangency portfolio on a Markowitz efficient line with an integer constraint. At the same time, the individual selection process comes with two downsides: first, the minimum costs come into effect for any asset in the portfolio; and second, unlike the fund, the individual portfolio does have integer and non-negativity constraints. Nonetheless, for the given market situation the advantages of individual portfolio selections outweigh these shortcomings in most cases – in particular when considering that investing in funds often comes at higher transaction costs than does purchasing shares.[12]

---

[12] The assumption of a perfect market for the market fund is based on the idea of having a "best case scenario" for the fund. In practice, however, funds usually do have a non-negativity constraint, and asset purchases often assume certain lot sizes which, too, affect the granularity of the assets' weights. The second of the mentioned disadvantages of individual portfolio selection is therefore less serious than assumed in this study, the actual advantages of individual stock selection might therefore be even greater.

3.5   Conclusions

In this chapter, a model for portfolio selection under fixed and/or proportional transaction costs together with non-negativity and integer constraints was presented and empirically studied on basis of DAX data. The major results from this study are that the presence of transaction costs might lead to significantly different results than for a perfect market and that the types of costs the investor occurs have different effects. Also, the optimal solution under transaction costs can not always be derived from the solution for frictionless markets.

Introducing transaction costs might have severe effects on the optimal portfolio structure. Even low fixed costs can lead to a substantial reduction in the number of different assets that ought to be included in a portfolio; the same is true for proportional costs or compound cost schemes. In due course, the asset weights might differ substantially. An investor facing transaction costs might therefore even have distinct advantages from individual stock selection over investing into a market fund – provided she does not simply try to replicate its weights but includes the relevant costs and additional constraints into the optimization process. Unlike claimed for a perfect market situation, in real world it might therefore be reasonable to hold a portfolio that deviates from the market portfolio even when all investors have homogeneous expectations. Another conclusion from these results is that investors might have advantages when they can invest in funds that are not just tracking the market portfolio, but also anticipate their clients' transaction costs.

Chapter 4

Diversification in Small Portfolios

4.1   Introduction

One of the central results of Modern Portfolio Theory is that, in perfect markets with no constraints on short selling and frictionless trading without transactions costs, investors will want to hold as many different assets as possible: Any additional security that is not a linear combination of the already included assets will contribute to the portfolio's diversification of risk and could therefore increase the investor's utility.

In practice, however, this situation is rather impractical, since the amount of transactions costs which has to be paid for many different small stocks, would raise the total cost considerably as has been shown in chapter 3. Moreover, the administration of such portfolios with a large number of different assets may become very tedious. Hence, investors seem to prefer portfolios with a rather small number of different assets (see, e.g., Blume and Friend (1975), Börsch-Supan and Eymann (2000), Guiso, Jappelli, and Terlizzese (1996) or Jansen and van Dijk (2002)).

Another important aspect in portfolio selection is that most of the risk diversification in a portfolio can be achieved with a rather small, yet well chosen set of assets.[1] Hence, in practice, the crucial question of finding the right weight for an asset is linked to the problem whether or not to include this asset in the first place.

Following Maringer (2001), Maringer (2002b), Keber and Maringer (2001) and Maringer and Kellerer (2003) as well as based on additional computational studies,

---

[1]   See, e.g., Elton, Gruber, Brown, and Goetzmann (2003, chapter 4) and section 1.1.1.

this and the next chapter are concerned with the portfolio optimization problem under cardinality constraints, i.e., when there is an explicit constraint on the number of different assets in the portfolio, with two alternative underlying models: First, the case of an investor is considered who, much in the sense of the previous chapter, wants to optimize her Sharpe Ratio in a modified Tobin framework, whereas in the next chapter a Markowitz framework will be assumed and the effects on efficient lines will be investigated. The remainder of this chapter is organized as follows. In section 4.2, the optimization problem and the optimization method will be presented. Section 4.3 summarizes the main results from an empirical study of this issue, section 4.4 concludes.

4.2   The Model

4.2.1   The Optimization Problem

From a theoretical point of view, the portfolio selection problem with a cardinality constraint can be regarded as *Knapsack Problem (KP)*[2]. The KP in its simplest version deals with selecting some of the available goods by maximizing the overall value of the resulting combination (objective function) without exceeding the capacity of the knapsack (constraint(s)). The investor's problem, however, demands two significant modifications of the classical KP:

- In the classical KP, each good has a given value which does not depend on what other goods are or are not in the knapsack. For portfolios, however, the "value" of any included asset depends on the overall structure of the portfolio and the other assets in the knapsack because of the diversification effects.

- In the classical KP, the goods have fixed weights, and one has to decide whether to take the good or not ("0/1 KP"). The investor has to jointly decide (i) whether to include an asset or not *and* (ii) what amount of her endowment to invest in this asset.

---

[2] See section 2.1.1 and Kellerer, Pferschy, and Pisinger (2004).

In this chapter we assume that investors can choose among one risk-free asset and up to N risky securities and want to maximize their (standardized) risk premium.

Tobin (1958) showed that any portfolio consisting of one risk-free asset and one or many risky assets will result in a linear relationship between expected return r and the risk associated with this portfolio, σ,

$$r = r_s + \theta_P \cdot \sigma$$

where r_s is the return of the safe asset. θ_P is the risk premium per unit risk[3] and is often referred to as *Sharpe Ratio, SR_P*.[4] Given the standard assumptions on capital markets with many risky and one risk-free asset, a rational risk averse investor will therefore split her endowment and invest a portion of α in the safe asset and $(1 - \alpha)$ in some portfolio of risky assets, $\mathcal{P}$, where the structure of $\mathcal{P}$ determines θ_P. The investor will therefore choose the weights, x_i, for assets i within the portfolio of risky assets in order to maximize the investment's risk premium per unit risk, θ_P. In passing note that the investor's level of risk aversion is reflected in her α and that the x_i's ought to be the same for any investor. Thus, the portfolio $\mathcal{P}$ (usually called *tangency portfolio*) can be determined without considering the investor's attitude towards risk and regardless of her utility function.

If there exists a market $\mathcal{M} = \{1, .., N\}$ with N assets k of which shall be included in the portfolio $\mathcal{P}$, the investor's optimization problem can be written as follows:

$$\max_{\mathcal{P}} \theta_P = SR_P = \frac{r_P - r_s}{\sigma_P}$$

subject to

$$r_P = \sum_{i=1}^{N} x_i \cdot r_i$$

$$\sigma_P = \sqrt{\sum_{i=1}^{N} \sum_{j=1}^{N} x_i \cdot x_j \cdot \sigma_{ij}}$$

---

[3] See also the presentation in section 1.1.2.3.

[4] As argued in section 1.2.1, the Sharpe Ratio was introduced as an *ex post* measure whereas θ is used in an *ex ante* optimization framework; meanwhile, this distinction is no longer made and the term "Sharpe Ratio" is used for *ex ante* considerations as well.

$$\sum_{i=1}^{N} x_i = 1 \quad \text{and} \quad \begin{cases} x_i \geq 0 & \forall i \in \mathcal{P} \\ x_i = 0 & \forall i \notin \mathcal{P} \end{cases}$$

$$\mathcal{P} \subset \mathcal{M}$$

$$|\mathcal{P}| = k$$

where σ_{ij} is the covariance between the returns of assets i and j and r_i is the return of asset i.

Like the (actually simpler) "0/1 KP" this optimization problem is NP–hard. It is usually approached by rules of the thumb (based on certain characteristics of the individual assets[5]) or by reducing the problem space by making *a priori* selections (e.g., by dividing the market into several segments and choosing "the best" asset of each segment[6]). As neither of these methods reliably excludes only "irrelevant" combinations, they tend to result in sub-optimal solutions. An alternative way to solve the problem is the use of meta-heuristics which are not based on *a priori* neglecting the majority of the problem space. The method suggested here has its origin in biology, namely ant systems.

4.2.2 Ant Systems

Biologists found that ants lay pheromone trails while foraging and carrying the found food to their nest. These trails serve themselves and their followers for orientation. Since shorter paths are likely to be traversed more frequently within the same period of time, these paths are also likely to have stronger pheromone trails – which will attract more ants that will then lay even more trails and so on. By this simple, reinforced strategy, ants are able to find shortest routes between two points, i.e., the nest and the food source.[7]

As presented in section 2.3.3, Dorigo, Maniezzo, and Colorni (1991), Colorni, Dorigo, and Maniezzo (1992a), Colorni, Dorigo, and Maniezzo (1992b) and Dorigo

---

[5] One popular rule, which will also be applied in this study, states to prefer assets that have high Sharpe Ratios SR_i.

[6] See, e.g., Farrell, Jr. (1997, chapter 4) on Asset Classes. Lacking appropriate information for our data sets, this approach could not be applied.

[7] See also section 2.3.3 for an illustrative description of the concept.

(1992) first introduced this principle to routing problems (such as the Traveling Salesman Problem[8]) by simulating real routes and distances between the cities in artificial nets and implementing an artificial ant colony where the ants repeatedly travel through these nets. Meanwhile, this concept resulted in the closely related meta-heuristics *Ant Systems* (*AS*) and *Ant Colony Optimization* (*ACO*) which have been applied successfully to a wide range of logistic problems and ordering tasks.[9] In particular, the introduction of *elitists* turned out to be very effective. In this concept the best solution found so far is reinforced each iteration in addition to the ants of the colony a certain number of elitist ants are traveling along the best solution found so far and by doing so reinforce this path.[10] In addition Bullnheimer, Hartl, and Strauss (1999) suggest a ranking system where ants with better solution spread more pheromone than the not so good ones and where paths of bad solutions receive no additional scent.

The concept of ant colony optimization and how it can be implemented will be presented in the next section. We will also demonstrate that this approach can be adopted for the Knapsack Problem in general and the investor's problem in particular.

4.2.3 The Algorithm

4.2.3.1 Approaching the Knapsack Problem

Applied to the portfolio selection problem, we implement an iterative search strategy where each iteration consists of three stages. In the first stage, artificial ants are to travel within a net consisting of N nodes which represent the available assets. An arc connecting any two nodes i and j where $i, j \in \mathcal{M}$ and $i \neq j$ shall capture whether a combination of these two is considered favorable or not. This can be achieved by introducing a matrix $[\tau_{ij}]_{N \times N}$ where τ_{ij} represents the amount of pheromone. The trail information will then be used to calculate the probabilities with which the following ants will select assets.

---

[8] See section 2.1.2.1.

[9] For a comprehensive survey on applications as well as the methodical variants, see, e.g., Dorigo and Di Caro (1999) or Bonabeau, Dorigo, and Theraulaz (1999).

[10] See Dorigo, Maniezzo, and Colorni (1996).

Let $\mathcal{P}'_a$ be the incomplete portfolio of ant a with $|\mathcal{P}'_a| < k$. If i is some asset already included in this portfolio, i.e., $i \in \mathcal{P}'_a$, whereas j is not, i.e., $j \notin \mathcal{P}'_a$, then the probability of choosing asset j shall be

$$p_{aj} = \begin{cases} \dfrac{\sum\limits_{i \in \mathcal{P}'_a} (\tau_{ij})^{\gamma} \cdot (\eta_{ij})^{\beta}}{\sum\limits_{i \in \mathcal{P}'_a} \sum\limits_{h \notin \mathcal{P}'_a} (\tau_{ih})^{\gamma} \cdot (\eta_{ih})^{\beta}} & \forall j \notin \mathcal{P}'_a \\ 0 & \forall j \in \mathcal{P}'_a \end{cases} . \tag{4.1}$$

The probability p_{aj} is mainly influenced by the amount of pheromone τ_{ij} that is on the paths from nodes $i \in \mathcal{P}'_a$ to node j. γ is a parameter for tuning that influence. In line with other implementations of ant based strategies, a matrix $[\eta_{ij}]_{N \times N}$ is introduced which represents the visibility of j from i. In routing problems, this information (which unlike the pheromone trails remains unchanged during the optimization) provides sort of a map thus providing the ants with guidelines or *a priori* information on preferred combinations. In the asset selection problem, $\eta_{ij} \geq 0$ might be used to indicate whether the investor regards combination i and j as desirable or not by transferring some general rule of the thumb onto $[\eta_{ij}]$. Also, the visibility could be employed to reinforce constraints. E.g., if i and j represent common and preferred stocks, respectively, of the same company and the investor does not want to hold both in her portfolio, she will set $\eta_{ij} = 0$, and the probability, asset j is added to a portfolio $\mathcal{P}'_a$ already containing i will become zero. If, on the other hand, she has a strong preference for this combination, a high value for η_{ij} will increase the probability that both i and j get included in the portfolio.

Results for the Traveling Salesman Problem suggest that in addition to elitists and ranking systems, the ants ought to be provided with some sort of a "road map" which usually is based on some *a priori* heuristics and is captured in the visibility matrix $[\eta_{ij}]$.[11] In ordering problems such as the Traveling Salesman Problem, the number of updated trails is rather small because the sequence in which the nodes are selected is of central importance. Thus, an ant visiting k nodes will update just $k - 1$ arcs and the chance of not updated arcs and evaporation on "good" arcs must not be neglected.

In our problem, however, it is the combination that matters, thus an ant selecting k securities will update $k \cdot (k - 1)$ arcs in a symmetric matrix. Having experimented

---

[11] See Bonabeau, Dorigo, and Theraulaz (1999).

with general rules and incorporated them into $[\eta_{ij}]$,[12] we found that they have a rather limited effect on the overall result: favorable parameters have been found to have far more influence on the reliability of the results and the speed at which the algorithm converges. We therefore do not introduce heuristics and "save" the visibility matrix for enhanced optimization problems, e.g., with possible individual preferences. In this study we assume that there are no such preferences and that the investor is interested only in maximizing the portfolio's risk premium per unit risk, θ_P. Thus, we set the visibility matrix $[\eta_{ij}] = 1$ and the parameter for tuning its influence $\beta = 1$. By also setting the parameter $\gamma = 1$, the probability from equation (4.1) melts down to

$$
p_{aj} = \begin{cases} \dfrac{\sum\limits_{i \in \mathcal{P}'_a} \tau_{ij}}{\sum\limits_{i \in \mathcal{P}'_a} \sum\limits_{h \notin \mathcal{P}'_a} \tau_{ih}} & \forall j \notin \mathcal{P}'_a \\ 0 & \forall j \in \mathcal{P}'_a \end{cases}.
\tag{4.1*}
$$

Once any ant has chosen their k assets, stage two of the model can be entered and the optimal portfolio weights are determined by some standard solution: When short sales are permitted, any ant can determine the maximum risk premium θ_{P_a} that can be achieved with the securities in its knapsack by the exact solution presented in equations (1.13) on page 13.[13] However, since our optimization model disallows negative asset weights, determining the θ_{P_a} is regarded as a quadratic programming problem as stated in section 2.1.2.4.[14]

In the third stage, when all ants have packed their knapsack and know their θ_P's, the trail information can be updated which comprises three steps:

- As time goes by pheromone evaporates. Thus, when a period of time has elapsed, only $\rho \cdot \tau_{ij}$ of the original trail is left where $\rho \in [0, 1]$.

---

[12] E.g., such rules could make use of the general result that diversification will be the higher the lower the correlation between the included assets. Hence, the visibility matrix could be derived from the correlation matrix or the covariance matrix by increasing (decreasing) the visibility between i and j when the correlation or covariance is low (high).

[13] See Keber and Maringer (2001).

[14] For alternative approaches, see, e.g., Elton, Gruber, Brown, and Goetzmann (2003) and Brandimarte (2002). In chapter 5, an algorithm will be presented that unites the two steps of asset selection and weight optimization.

- New pheromone trails are laid. In real life, ants tend to be permanently on the run and are permanently leaving new pheromone trails without bothering whether their colleagues have already returned to the nest. In artificial ant systems, however, each ant chooses one path through the net and waits for the other ants to complete their journey before starting the next trip. Thus, artificial ant systems usually assume that any ant a spreads a fixed quantity Q of pheromone on its path which has a length of L_a and by doing so updates the trail by $\Delta_a \tau_{ij} = Q/L_a$ for any arc (i, j) along a's path. This implies that the shorter the path the higher the additional trail. Since both in real life and in ant systems L_a is to be minimized whereas here θ_{P_a} is to be maximized, we adopt this concept and use $1/\theta_{P_a}$ for a substitute of L_a. Thus, the trail update for ant a would be $\Delta_a \tau_{ij} = Q \cdot \theta_{P_a}$ for all securities $i \neq j$ and $i, j \in \mathcal{P}_a$.

Bullnheimer, Hartl, and Strauss (1999) suggest a ranking system that reinforces the solutions of the best ants of the current population (here called "prodigies"). In our application, this concept allows only the w best ants to update $[\tau_{ij}]$ where the rank $\mu = 1, .., w$ determines the quantity of pheromone Q_μ a prodigy can spread. Assuming a simple linear ranking system where the quantity of pheromone depends on the ant's rank, prodigy μ updates arc (i, j) by

$$\Delta \tau_{ij,\mu} = \begin{cases} Q_\mu \cdot \theta_{\mathcal{P}^\mu} & \forall i, j \in \mathcal{P}^\mu, i \neq j \\ 0 & \text{otherwise} \end{cases} \qquad (4.2)$$

where

$$Q_\mu = \begin{cases} ((w - \mu) + 1) \cdot Q & \mu \leq w \\ 0 & \mu > w \end{cases}.$$

- Assuming that, in addition to the ants of the current colony, ε elitist ants are choosing the best portfolio found so far, $\mathcal{P}^*$, and each of them spread Q pheromone, then each elitist updates the matrix by

$$\Delta \tau_{ij}^* = \begin{cases} Q \cdot \theta_{\mathcal{P}^*} & \forall i, j \in \mathcal{P}^*, i \neq j \\ 0 & \text{otherwise} \end{cases}.$$

```
initialize pheromone matrix τ_ij = τ⁰∀i ≠ j and τ_ii = 0;
population size := N;
REPEAT
    FOR a := 1 TO Population size DO
        P_a := {a};
        WHILE |P_a| < k DO
            determine selection probabilities p_aj∀j ∉ P_a according to
                definition (4.1*);
            use probabilities p_aj to randomly draw one additional asset j;
            add asset j to the portfolio, P_a := P_a∪{j};
        END;
        determine optimal asset weights such that {SR|P_a} → max!;
    END;
    rank ants according to their portfolios' Sharpe Ratios;
    IF max SR_{P_a} > SR_{P*}
        new elitist is found, replace previous elitist P* with new one;
    END;
    update pheromone matrix;
UNTIL convergence criterion met;
REPORT elitist;
```

Listing 4.1: Pseudo-code for the main Ant System routine

Combining evaporation and new trails, the pheromone matrix is to be updated according to

$$\tau_{ij} := \rho \cdot \tau_{ij} + \sum_{\mu=1}^{\omega} \Delta\tau_{ij,\mu} + \varepsilon \cdot \Delta\tau_{ij}^* \quad \forall i \neq j. \tag{4.3}$$

Due to this updates, the next troop of ants can apply their predecessors' experiences: In the next iteration, the probabilities according to (4.1*) will be influenced, and the ants' preferences will be shifted towards combinations of securities that have proven successful. Listing 4.1 summarizes the main steps of the algorithm. The computationally most complex parts of the algorithm are the computation of a portfolio's volatility, $\mathcal{O}(k^2)$, having to sort the population, $\mathcal{O}(A \cdot \ln(A))$ where A is the number of ants in the colony, and the update of the pheromone matrix which is quadratic in k and linear in the number of prodigies plus the elitist, $\mathcal{O}((k^2 - k) \cdot (\omega + 1))$, and quadratic in N due to the evaporation, $\mathcal{O}(N^2)$. The overall complexity of the algorithm is determined by these individual complexities times the number of iterations.

4.2.3.2   First Applications of the Algorithm

In order to determine the essential parameters for the algorithm we ran a series of tests with random numbers for elitists, ε, prodigies, ω, and factor of evaporation, ρ. We then tried to find correlations between these values and the "effectiveness" (i.e., speed and level of improvement during the iterations) of the respective ant colony.[15] According to these results we chose the colony size to equal the number of available assets, i.e., 100, $\varepsilon = 100$ (i.e., equal to the number of securities and ants per iteration), $\omega = 15$ (i.e., only the best 15 per cent of ants where allowed to update according to equation (4.2)), and $\rho = 0.5$ (i.e., half of last round's trail information evaporates).

The version of the ant algorithm as just presented was first applied to the portfolio selection problem in Maringer (2001) where the *ex ante* Sharpe Ratio (SR) (which, as argued, is equal to θ_P) is to be maximized for a DAX data set. In Maringer (2002b) it is applied to finding cardinality constrained portfolios under a Markowitz/Black framework, and the computational study therein is based on subsets of a FTSE data set with $N = 25, \ldots, 90$ where the expected return is equal to the market's expected return and the risk is to be minimized, and the algorithm is found to be superior to Simulated Annealing and a Monte Carlo approach. In addition, for a number of problems, the heuristically obtained results were compared to those from complete enumeration, and it was found that the ant algorithm reported the optimal solution in the majority of runs (often all the runs) even when the alternative method Simulated Annealing was unable to identify the optimum even once. In Keber and Maringer (2001), the ant algorithm is compared to Simulated Annealing and Genetic Algorithm based on the SR maximization problem with the same FTSE data set. Again, it was found that the results of population based heuristics are superior to Simulated Annealing, that, however, the population based heuristics also take more time to find appropriate parameter values.

For the sake of simplicity neither of these studies had a non-negativity constraint on the weights which is included in this study. In addition, in all three previous studies, the ant algorithm did exhibit a typical property of this method: Ant algorithms perform best when applied to large problems, whereas it is the rather "small" problems sometimes that cause slightly more problems to the ant algorithm. When k, the

---

[15] For a short presentation to the parameter selection problem, see section 2.4 and Winker (2001, chapter 7).

number of different asset included in the portfolio, is rather small, the algorithm converges quite fast, and the colony might get stuck in a local optimum which they cannot escape. In Maringer (2002b), e.g., selecting $k = 3$ assets appears more demanding than selecting $k = 6$ assets for all markets with $N \leq 78$. Though even for these cases, the chances of identifying the actual global optimum in an independent run is for the ant algorithms by magnitude higher than the results found by Simulated Annealing, it is desirable to have equally high reliability for small problems.

4.2.3.3   Refining the Algorithm

In heuristic search strategies, a common stopping criterion is the number of iterations the elitist has not changed, i.e., for how many iterations the algorithm has produced no further improvement. The search is then stopped and the current elitist is reported. If there is a chance that the result is only a local one, a new independent run is started and knowledge acquired in the previous run is lost. In Artificial Intelligence, another common way of overcoming a potential local optimum is to introduce a random shock:[16] When an agent has not achieved an improvement over a given number of iterations, it is randomly "positioned" at a different location and will continue the search from there, yet without necessarily starting a new, perfectly independent run. Based on this idea, we suggest a similar concept to the ant algorithm.

In ant algorithms (as well as in real life), the ants tend to get stuck in a local optimum when the pheromone trails for a good, yet not globally optimal solution are so strong that chances of finding a route aside these tracks are very low – and are even lowered in due course as the (probably) suboptimal routes are reinforced. Enforcing alternative routes therefore demands lowering the pheromone level on these (probably) suboptimal tracks. We suggest a simple means that might do exactly this trick: With a certain probability the initial pheromone matrix (or a weighted combination of the current and initial matrices) is reinstated yet the current elitist is kept. This implies that knowledge and experience acquired in previous runs is kept while the ants have a higher chance of selecting alternative routes. Metaphorically speaking, this corresponds to "rain" where current pheromone trails are washed away or at

---

[16]   See, e.g., Russell and Norvig (2003).

least blurred. We therefore introduce a reset parameter $v \in [0,1]$ where $v = 1$ corresponds to "heavy rain" where all pheromone trails are swept away and the original pheromone matrix is restored; the closer v is to zero, the more of the current trail information endures. This variant has a similar effect as the local updates of the pheromone trail in *Ant Colony Optimization (ACO)*,[17] a variant of ant algorithms which is to foster diversity within the colony's solutions.

The updating rule (4.3) for the off-diagonal elements of the pheromone matrix, $[\tau_{ij}]$ with $i \neq j$, can then be enhanced with an option where a part of the old and newly added trails are washed away and the initials trails are restored:

$$
\tau_{ij} := \begin{cases} \left(\rho \cdot \tau_{ij} + \sum_{\mu} \Delta \tau_{ij,\mu} + \varepsilon \cdot \Delta \tau_{ij}^* \right) \cdot (1 - v) + \tau^0 \cdot v & \text{"rain"} \\ \rho \cdot \tau_{ij} + \sum_{\mu=1}^{\omega} \Delta \tau_{ij,\mu} + \varepsilon \cdot \Delta \tau_{ij}^* & \text{"sunshine"} \end{cases} \quad (4.3^*)
$$

where the option "rain" is chosen with a probability of p_{rain} and the alternatively chosen option "sunshine" corresponds to the original updating rule (4.3). Whether this concept is advantageous or not, was tested in a computational study; the main results will be discussed in section 4.3.2.

4.3   The Empirical Study

4.3.1   The Data

The empirical study in this chapter is based on data sets for the DAX, the FTSE, and the S&P 100. The *DAX data set* contains the 30 stocks represented in the German stock index DAX30. The *FTSE data set* is based on the 100 stocks contained in the London FTSE100 stock index; four of the stocks, however, had to be excluded due to missing data. In both cases we used daily quotes over the period July 1998 – December 2000. Based on the corresponding historic returns we calculated the covariances σ_{ij} which are used for estimators of future risk. The expected returns, r_i, were generated with a standard *Capital Asset Pricing Model (CAPM)* approach[18] according to $r_i = r_s + (r_M - r_s) \cdot \beta_i$ with an expected safe return of $r_s = 5\%$, expected market

---

[17] See Bonabeau, Dorigo, and Theraulaz (1999, p. 49) and the literature quoted therein.

[18] See section 1.2.2.

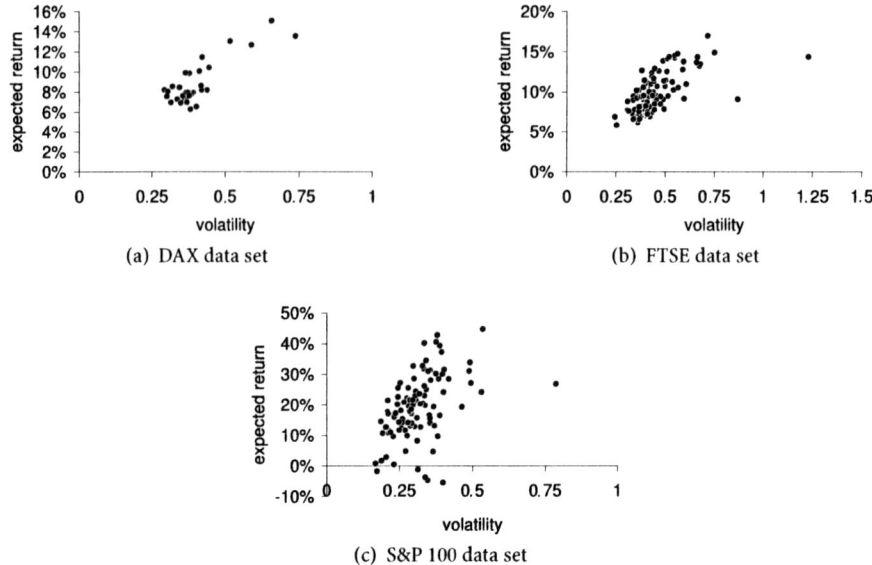

Fig. 4.1: Estimated return and risk for the data sets

risk premia of $r_{DAX} - r_s = 5.5\%$ and $r_{FTSE} - r_s = 6\%$, respectively,[19] and with beta coefficients, β_i, coming from the historic returns. The distributions of the assets in the return-volatility space are depicted in Figures 4.1(a) and 4.1(b), respectively. In the light of recent developments in the capital markets, we want to point out that we focus exclusively on the selection problem and that in this optimization problem the mean and variance of returns are regarded as exogenously determined.

Estimating the assets' returns via the CAPM implies that their (estimated) Sharpe Ratio differ in their correlation with the market:

$$SR_i = \frac{\overbrace{(r_s + (r_M - r_s) \cdot \beta_i)}^{=r_i} - r_s}{\sigma_i}$$
$$= \frac{(r_M - r_s) \cdot \frac{\sigma_i \cdot \rho_{iM}}{\sigma_M}}{\sigma_i}$$

---

[19] As in chapter 3, the values for the safe interest rate and the markets' risk premia were chosen to represent what then would have made reasonable guesses. With the focus on the optimization where the estimates for risk and return can be considered exogenously determined, the actual values proofed to have little influence on the conclusions drawn in the computational study.

$$= \frac{r_M - r_s}{\sigma_M} \cdot \rho_{iM}$$
$$= SR_M \cdot \rho_{iM}$$

where $\mathcal{M}$ is the respective market index. Though this does not affect the covariances of any two assets and therefore does not have an immediate effect on a portfolio's volatility and Sharpe Ratio, the values for the third data set, the *S&P 100 data set*, are estimated differently. Based on daily returns for the stocks in the S&P 100 stock index from November 1995 through November 2000 and for 23 country, 42 regional and 38 sector indices, the expected returns for the stocks where estimated from the first 1 000 days with an combined APT[20] and GARCH[21] approach: First, for any asset the bundle of five indices was determined that explains most of the asset's return in sample. Next, the expected returns and volatility for the indices where estimated with a GARCH model and the assets' expected out of sample returns based on the individual APT models. The assets' volatilities where estimated with a GARCH model, the covariances where determined with the assets' volatilities and their historic correlation coefficients. The factor selection process is presented in chapter 7 which also offers a more detailed presentation of the underlying method. The results from this estimation procedure appear quite reliable: Only for eight of the 100 assets, the actual out of sample returns differ statistically significant from their expected values.[22] The volatilities and expected returns for this data set are depicted in Figure 4.1(c). Though comparable to the DAX data set in the range of the assets' volatilities and the FTSE data set in the number of assets, the S&P data set differs from the others in the range of the expected returns since it also contains assets with negative expected returns.[23]

---

[20] See Ross (1976) and section 1.2.4. A detailed presentation of the data set and the factor selection problem can be found in chapter 7.

[21] See Engle (1982), Bollerslev (1986) and Engle (1995), and the presentation in sections 1.1.3.3 and 2.4.

[22] Significance test at the usual 5% level; corresponding tests for the DAX and FTSE data sets had to be omitted in lack of out of sample data. The following presentation will therefore focus strictly on the selection problem given a certain market situation which can be considered realistic.

[23] When assets are negatively correlated with the index, both the APT and the CAPM predict negative risk premia in equilibrium which, when exceeding the safe interest rate, can also result in negative expected returns. The economic argument behind is that an investor is willing to pay an ("insurance") premium for assets that react opposite to the market trend and are therefore well suited for diversification and hedging.

4.3.2   Computational Study for the Modified Update Rule

In order to test whether the concept of casually resetting the pheromone matrix has a favorable effect on the algorithm's performance or not, we ran a series of Monte Carlo experiments where in the initialization step the value of the reset parameter ν was randomly chosen from $(0.1, 0.25, 0.50, 0.75)$ and the probability for "rain" was chosen to be $p_{rain} \in \{5\%, 10\%, 25\%, 50\%, 75\%\}$, i.e., with a probability of p_{rain} of the update steps, the modified update rule (labeled "rain" in (4.3*) was applied and with a probability of $(1 - p_{rain})$, the original update rule (4.3) (labeled "sunshine" in (4.3*)) was applied. For each combination of parameters and different value of assets in the portfolio, k, approximately 120 independent runs were performed. For comparison, we also ran the algorithm in its previous version with update rule (4.3) without the "rain" modification by simply setting $p_{rain} = 0\%$; here 1 000 independent runs were performed. For any parameter setting, the number of iterations per run was limited to 200, the colony size equaled the number of included assets, k.

Based on the S&P 100 data set, the two cases $k = 3$ and $k = 10$ are considered. These two problems differ considerable in the number of candidate solutions: the former comes with just 161 700 alternatives,[24] the latter with 1.73×10^{13} alternatives.

Figure 4.2 depicts the range for the reported solutions depending on the different values for ν and p_{rain}. As can be seen for either value of k, the version without rain reports quite diverse solutions. Though the global optimum is found eventually, a high number of runs is necessary to reduce the likelihood that just a local optimum is reported: for $k = 3$, in just 16% of all runs, the global optimum was found, and for $k = 10$, in just 2 of the 1 000 independent runs the global optimum was reported.

With appropriate values for p_{rain} and ν, on the other hand, the algorithm performs significantly better: For the case with $k = 3$, in two thirds of the runs the global optimum was identified by any of the tested version with $p_{rain} \in (5\%, 10\%, 25\%)$ (and arbitrary positive value for ν) or with $\nu = 0.10$ (and arbitrary positive values for p_{rain}). For the case where $k = 10$, in two thirds of runs the global optimum

---

[24]   From a practical point of view, for $k = 3$, the number of alternatives would be small enough for an exhaustive search. Nonetheless, a heuristic optimization method ought to work well with small problems; hence this case is considered here, too.

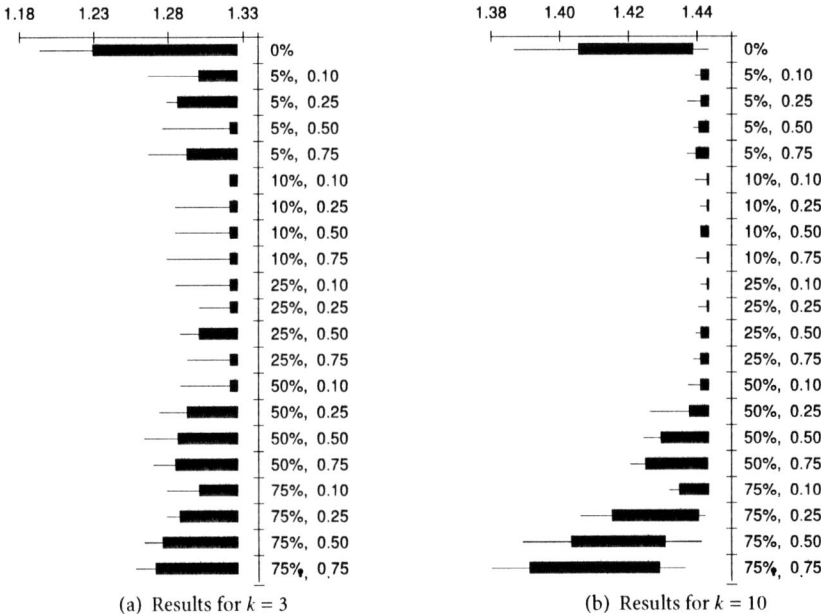

Fig. 4.2: Range from best to worst reported result (lines) and range covering 90% of reported solutions (black area) with the traditional update rule (4.3) (with p_{rain} = 0%) and with rain according to rule (4.3*) for different values of p_{rain} (in %) and v (as decimals)

was reported with the parameter combinations $(p, v) = (5\%, 0.10)$ and $(10\%, 0.25)$, and for any positive value of v with $p_{rain} \in (5\%, 10\%, 25\%)$, half of the runs reported the global optimum.

Additional experiments showed that the performance of the traditional version without rain could be improved by increasing the number of iterations, yet never reached the modified version's high ratio of runs in the global optimum was identified.

4.3.3   Financial Results

According to the theory, increasing the number of assets k in the portfolio $\mathcal{P}$ causes an increase in the risk premium $SR_{\mathcal{P}}$ provided the right assets are chosen and assigned the optimal weights. In addition, the marginal contribution of any additional security to the portfolio's diversification is decreasing. Both effects can be found in the results for either of the markets: The graphs in Figure 4.3 depict the bandwidth within which the Sharpe Ratio will lie when the asset weights are optimized for the best and the worst possible combination of k assets.[25] As can easily be seen, in all three markets, a relatively small yet well chosen set of assets can achieve a higher $SR_{\mathcal{P}}$ than a large yet badly chosen set of assets. E.g., in the FTSE data set, the optimal combination of $k = 5$ assets might outperform a poor combination of $k = 84$ assets: $SR_{\mathcal{P}(k=5)}^{\max} = 0.2863 > SR_{\mathcal{P}(k=84)}^{\min} = 0.2860$. Note that this is all due to good or bad selection of the assets and not due to suboptimal asset weights.[26]

The bandwidth for the Sharpe Ratio will be the larger the smaller the portfolio and the larger and more diverse the market is: selecting any $k = 15$ assets from the DAX data set, e.g., the achievable SR will be in the range from 0.1658 to 0.2447; if the same number of different assets is selected from the FTSE and the very diverse S&P data set, the Sharpe Ratios will range from 0.1294 to 0.3202 and from 0.2545 to 1.4497, respectively.

---

[25] Finding the "worst" combinations, too, represents an optimization problem where the sign in the objective function is changed, yet not the way the weights $x_i \forall i \in \mathcal{P}$ are determined. This assures that low values for the Sharpe Ratios actually do come from selecting the "wrong" assets and not by an inappropriate loading of the weights.

[26] See the approach of Solnik (1973) and its presentation in section 1.1.1.

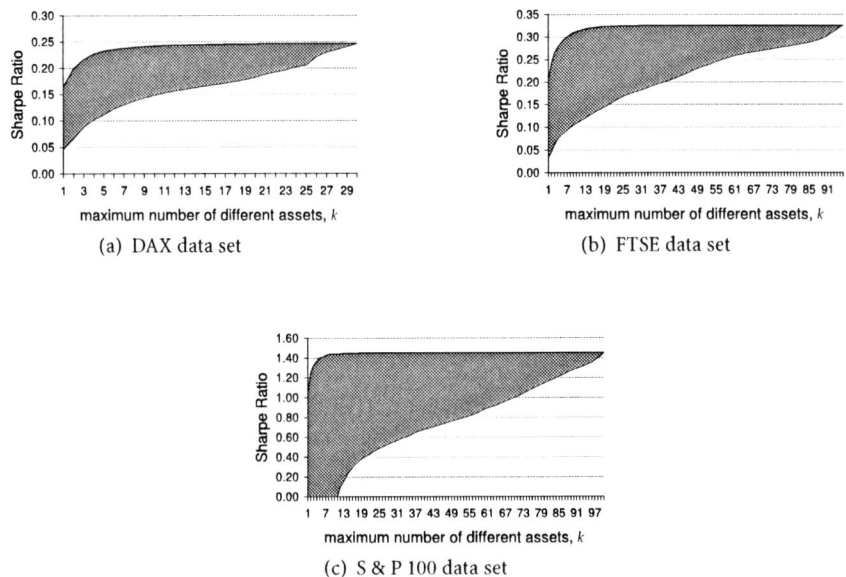

(a) DAX data set   (b) FTSE data set

(c) S & P 100 data set

Fig. 4.3: Range for Sharpe Ratios for portfolios under cardinality constraint with optimized weights

A simple rule of the thumb suggests to prefer assets which themselves have a high Sharpe Ratio. According to this rule, the available assets are sorted by their SR in descending order, and the first k assets are selected for the portfolio. A downside of this rule is that it does not consider the correlation or covariance between the assets which largely affects the portfolio's volatility. Hence, this rule will not necessarily find the optimal solution, particularly when k is rather small. Having a method that has a higher chance of identifying the actually best combination, one can also evaluate how large the gap between the SR rule based portfolios and the optima is. As can be seen from Figure 4.4 for the DAX data, the Sharpe Ratios of portfolios selected with this popular rule could mostly be achieved with one or more assets less. For the FTSE data set, this is even more apparent: For the optimal portfolio with $k = 10$, the SR is higher than for a portfolio with $k = 38$ when selected with the SR rule. The consequences of this gap become even more severe when the investor faces transactions costs that contain a fixed fee as can be seen from the results in the previous chap-

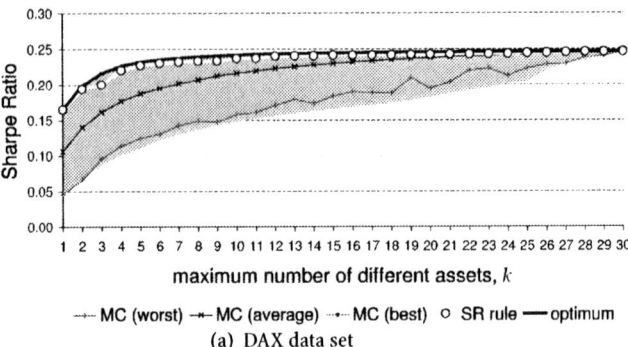

(a) DAX data set

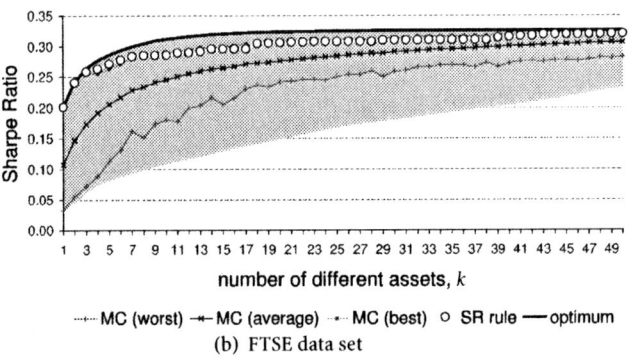

(b) FTSE data set

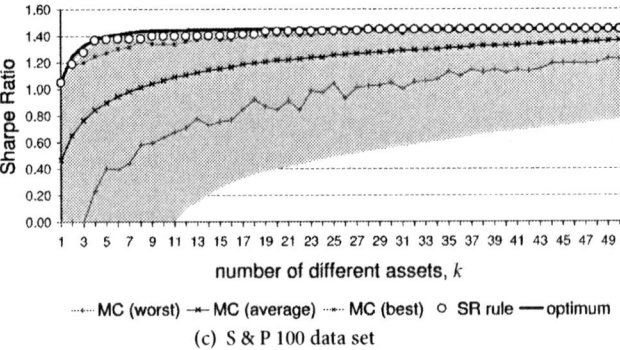

(c) S & P 100 data set

Fig. 4.4: Sharpe Ratios for portfolios under cardinality constraint with different selection processes

ter. Other rule-based selection methods such as selections based on the companies' industry, size, or geographic aspects exhibit equal shortcomings.

For the Monte Carlo approach, k assets are drawn randomly and their weights are optimized. This selection process has been replicated a 1 000 times, and the best, the worst, and the average SR for any k and data set are plotted on the bandwidth for the possible outcomes. As can be seen for the larger markets, it is very unlikely to randomly draw the worst possible solution – yet it is also unlikely that the optimal solution is chosen: In the best of the 1 000 replications, a solution close to that from the SR rule is found, on average, however, a random selection is significantly below what could be achieved with a superior selection method: the upper limit, indicating the optima are the results found with the heuristic search method.

As neither the SR rule nor the MC approach includes the correlations and co-variances between the assets into the selection process, a main aspect from portfolio selection might be lost. A closer look at what assets actually are selected and what weights they are given also confirms that the decision of whether to include a certain asset or not depends on what other assets are included. In Figure 4.5 the cumulated asset weights are depicted for the different values of k. In particular the results for the FTSE data set illustrate that the optimal selection with k assets cannot be determined by simply searching the asset that fits best to the solution with $k - 1$ assets: as has already been argued in the previous chapter, in smaller portfolios one asset might serve as a substitute for a bundle of other assets which, however, cannot be included because of the constraints (be it transactions costs, be it cardinality). Also, what makes a good choice in a portfolio with few different assets might or might not be a good choice for large portfolios.

The results for the S&P 100 data set (Figure 4.5(c)) also exhibits a particularity of this data set: Given the estimates for the assets' returns and covariances, only a limited number of assets are actually assigned positive weights, i.e., even for large k only a small number of different assets is included in the portfolio, and the cardinality constraint is no longer binding. The selection of these assets depends to some extent on the choice of the safe interest rate, r_s, as (geometrically speaking) the tangency line from the Tobin efficiency line crosses the y-axis of the mean-variance-diagram at a different point, yet the basic results are unchanged. The SR rule "filters" most of the assets that ought not to be included in any of the optimal portfolios but again still ignores some better combinations in the lack of considering the covariances. In

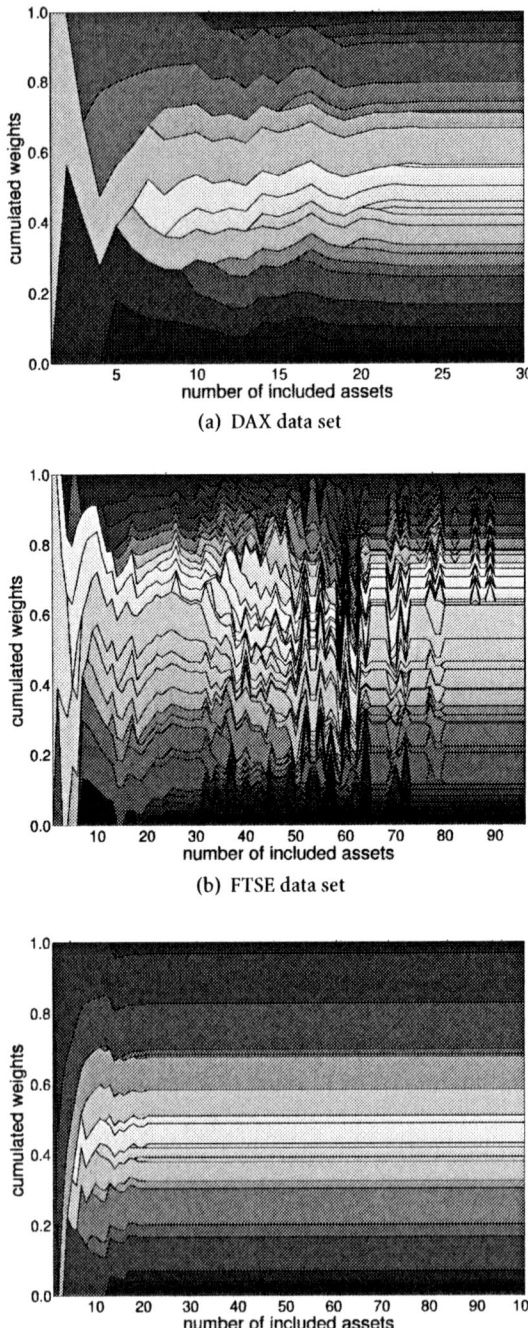

(a) DAX data set

(b) FTSE data set

(c) S & P 100 data set

Fig. 4.5: Cumulated weights for optimal portfolios under cardinality constraints

this type of market situation, however, a Monte Carlo approach will be likely to also pick one or several of these undesirable assets – and will therefore be clearly inferior to a heuristic search strategy.

4.4   Conclusion

For various reasons, investors tend to hold a rather small number of assets. In this chapter, a method has been presented to approach the associated NP hard optimization problem of selecting the optimal set of assets under a given market situation and expectations. The main results from this empirical study are twofold: (i) the well known fact of decreasing marginal contribution to diversification is not only confirmed, but can be exploited by identifying those assets that, in combination, offer the highest risk premium; (ii) it has been shown that alternative rules, frequently found in practice, are likely to underperform as they offer solutions with risk premia lower than would be possible under the same constraints and market situations.

Chapter 5

Cardinality Constraints for Markowitz Efficient Lines

5.1 Introduction

5.1.1 The Optimization Problem

A salient feature of financial portfolios is that any additionally included assets might contribute to the diversification of risk without necessarily decreasing the expected return. Hence, an investor seeking an optimal ratio between risk and risk premium within a Markowitz framework will seek to include as many different assets as possible.[1] The results in chapter 3, however, have shown that in non-perfect markets, there might be good reasons for not including all available assets even when the market is in equilibrium and all assets are fairly priced and there are no restrictions on the asset prices. In addition, it was found in chapter 4 that most of the diversification can be achieved with an even smaller number of different assets than is usually argued in the literature. Furthermore, it was argued that there exist various other grounds as well as empirical findings that investors prefer portfolios with a rather limited number of different assets.

Following these arguments, this chapter, too, focuses on the case that an investor wants to hold at most k different assets in her portfolio. The previous two chapters assumed that there exists a risk-free asset, implying that, by the separation theorem, the optimal weights of the risky portfolio can be found without explicit knowledge

---

[1] See section 1.1.2

of the (risk averse) investor's attitude towards risk.[2] In practice, however, a truly safe return, valid for any investor, is not always available (in particular when long investment horizons are considered). Consequently, the efficient line in the tradition of Markowitz has to be identified from which the investor will choose individually. This selection problem will be considered in this chapter.

If at most k out of the N available assets are to be included in the portfolio, then a cardinality constraint has to be introduced, e.g., by defining a binary variable b_i that is set to 1 when asset i is included and to 0 otherwise where the sum of all b_i's must not exceed k. In order to calculate the entire efficient frontier rather than the risk minimizing combination for a single given return, an exogenous parameter $\lambda \in [0, 1]$ is introduced. In correspondence to model (1.7) on page 7 with the modified objective function (1.7a*), this parameter is to regulate the trade-off between risk and return in the objective function by multiplying the expected return, r_P, with λ and the punishment term for the risk, $-\sigma_P$, with $(1 - \lambda)$: If λ is equal to or close to zero, then the risk term gets all or most of the weight, and portfolios from the efficient set with low volatility will be identified. The larger a value for λ is chosen, the more the weight is shifted towards the portfolio's expected return, and the optimization process will preferably identify those portfolios from the efficient set that have high expected yield.

Summarizing, the portfolio problem with cardinality constraint reads as follows:

$$\max \left(\lambda \cdot r_P - (1 - \lambda) \cdot \sigma_P \right)$$

subject to

$$r_P = \sum_{i=1}^{N} x_i \cdot r_i$$

$$\sigma_P = \sqrt{\sum_{i=1}^{N} \sum_{j=1}^{N} x_i \cdot x_j \cdot \sigma_{ij}}$$

$$x_i \geq 0 \quad \forall i$$

---

[2] See section 1.1.2.3.

$$\sum_{i=1}^{N} x_i = 1$$

$$\sum_{i=1}^{N} b_i \leq k \text{ where } b_i = \begin{cases} 1 & x_i > 0 \\ 0 & \text{otherwise} \end{cases}.$$

The optimal traditional algorithms which are well-known from the literature are based on existing algorithms for solving mixed-integer nonlinear programs,[3] and are thus not applicable to problems with a large number of variables: A common approach are "branch and bound methods"[4] which in the course of searching the solution demand a relaxation of constraints; difficulties can therefore emerge as the algorithm might eventually end up with some solution that is either infeasible, or that is just a local optimum that cannot be overcome because of the constraints. An alternative approach has therefore to be considered.

5.1.2 The Problem of Optimization

If there are N assets to choose from but only k are to be included in a portfolio, then there are $\binom{N}{k}$ alternatives. If an investor wants to have ten different stocks in her portfolio, selecting from the 4 200 stocks at the Frankfurt stock exchange comes with some 5×10^{29} alternatives – a number which corresponds to about 50 times the diameter of the universe when measured in millimeters. Reducing the set of available assets to 300 (which is approximately the number of shares quoted at the Hanover stock exchange) still offers one billion billions of different combinations, and restricting oneself to the fifty stocks contained in the Euro Stoxx 50 still comes with about ten billon possible combinations – the weight structure of which had still to be optimized. Finding the optimal solution by enumerating all the possible solutions is therefore out of the question.

To come to grips with this vast problem size, theory and practice offer different approaches. A popular way is to generate so called *Asset Classes* by grouping the available assets according to predefined criteria (such as industry, size, geographical aspects, and so on) and then preselecting from each of these groups one or a small

---

[3] See, e.g., Bienstock (1996) and Brandimarte (2002, chapter 6).

[4] See section 2.1.2.3.

number of assets which is/are considered to be the best within their groups.[5] The actual optimization is then performed on this reduced problem. The downside of this approach is that it has to *a priori* exclude the vast majority of potential solutions, and by doing so the actually optimal solution, too, might get dismissed. By selecting just the best asset within a group without considering how well it can be combined with the other selected assets, the salient feature of portfolios, the assets' contribution to the overall diversification of risk, has to be largely ignored. This corresponds well to the results from chapter 4 where it was shown that asset selection based on a rule of the thumb that ignores the actual correlations between assets can lead to severely inferior solutions.

In the previous and this chapter, two cases of the portfolio selection problem with a cardinality constraint are investigated: identifying the portfolio that has the highest Sharpe Ratio, and identifying the whole of the efficient set. For reasonable values of N, neither case can be solved with traditional optimization methods due to the complexity behind the problem, yet the two cases differ in one significant aspect: given a bundle of selected assets, standard optimization methods can be used to find the risk minimizing assets' weights, the heuristic can therefore focus on the selection problem. If, however, the whole efficient set has to be identified, then it appears favorable to have the selection and the weight allocation problems both solved by the heuristic.

The cardinality constraint makes the solution space quite rough and demanding. This type of problem could be approached with a single agent local search algorithms such as Simulated Annealing (SA) applied in chapter 3 or Threshold Accepting[6] (TA) – yet with a caveat: Both SA and TA can be shown to find the global optimum[7] given the heuristics' parameter meet certain requirements. In practice, determining the heuristics' parameters has to account for convergence quality as well as convergence speed, hence the relevant parameters, such as initial temperature and cooling factor for SA and threshold sequence for TA, respectively, the maximum number of iterations, definitions of local neighborhood, etc. are chosen in a way that are likely to find good solutions within reasonable time at the risk of getting stuck in a local optimum. To reduce the peril of eventually reporting such a local

---

[5] See Farrell, Jr. (1997) and section 1.3.

[6] See, e.g., Winker (2001).

[7] See section 2.3.1.

optimum, the optimization problem is usually solved several times in independent runs, and only the best of all results is considered.[8] Though acceptable for many optimization problems, it might become ineffective for solution spaces that are highly complex: The number of independent runs had to be increased noticeably, and experience gained in preceding runs is usually not reused. In addition and foremost, the concept of finding the global optimum by local search might demand to traverse the single agent the whole of the solution space when the starting point is chosen unlucky.

Hence, it is found that problems with many local optima and a highly complex solution space can be approached more efficiently with multi-agent methods that also incorporate global search aspects rather than with single-agent local search methods. These methods usually have elements that allow for pooling experience and passing on knowledge between the agents, and the available computational time can therefore be used more economically. The downsides of these methods are that their implementation is usually more demanding and fine-tuning of the involved parameters can become quite cumbersome.

In Maringer (2002b), a Black portfolio framework is enhanced with a cardinality constraint. This optimization problem is considered as a special version of the Knapsack Problem and then solved with Simulated Annealing and a modified version of Ant Systems.[9] The results show that a population based multi-agent is better suited to solve this optimization problem than the single-agent method SA. Equally, Keber and Maringer (2001) investigate an investor who wants to maximize her portfolio's Sharpe Ratio in a traditional Tobin framework as presented in chapter 3, yet enhanced with a cardinality constraint. Comparing Genetic Algorithms, the modified version of the Ant Systems and Simulated Annealing, all three methods are found capable of solving the problem, yet again the single-agent method SA is clearly outperformed by the two multi-agent methods.[10]

Though multi-agent global search methods, such as Ant Systems, Genetic Algorithms or Evolutionary Computing, usually exhibit a better convergence behavior, the lack of local search aspects in these methods might lead to reporting a result

---

[8] See also section 2.4.

[9] See also sections 2.1.1, 2.3.1 and 2.3.3 as well as chapter 4.

[10] See also Chang, Meade, Beasley, and Sharaiha (2000).

near the global optimum rather than the global optimum itself. It appears therefore desirable to have a method that combines global and local search elements. Based on Maringer and Kellerer (2003), this chapter will present such a method for the portfolio optimization problem with cardinality constraints. After formalizing the optimization problem, this hybrid algorithm is presented in section 5.2. Section 5.3 summarizes a comparative computational study where Simulated Annealing, a special version of Simulated Annealing and this algorithm are applied to data of stocks represented in the DAX and FTSE stock indices, respectively. In section 5.4 the implications of the cardinality constraints from the financial point of view are discussed, and section 5.5 concludes.

5.2   A Hybrid Local Search Algorithm

5.2.1   The Algorithm

In order to solve the presented problem we apply an iterative algorithm in which a "population" of crystals is to find the optimal portfolio structure. Each of these crystals represents a portfolio where the structure of a crystal depicts both the assets included and their respective weights. The algorithm starts with a random initialization of the crystals the structure of which is random yet valid with respect to the constraints. This is done by selecting k of the N available assets and assigning them random positive weights such that they add up to 1, i.e., $\sum_i x_{ict} = 1$ where x_{ict} is the weight of asset i in crystal c's portfolio in iteration $t = 0$.

The subsequent iterations consist of three stages: *modification* of each crystal's portfolio structure; *valuation* and ranking of the modified crystals; and *replacement* of the poorest crystals in the population. Reinforcement of promising portfolio structures takes place not only in the third stage where the weakest individuals are eliminated and replaced with supposedly stronger ones, but also in the first stage when assets are to be exchanged.

In iteration t the three stages comprise the following activities:

Modification. For each crystal c, d assets of this crystal's current portfolio, $\mathcal{P}_{ct}$, are selected. For these assets the respective weights are changed according to

$x'_{ict} = \max\{x_{ict} + \tilde{z}_{ict}, 0\}$ where $\tilde{z}_{ict} \in [-U_t; U_t]$ is an equally distributed random variable. The other assets' weights are left unchanged, i.e., $x'_{ict} = x_{ict}$. U_t indicates the bandwidth for changes in iteration t which is steadily narrowed, $U_t = U_{t-1} \cdot \gamma_U$ with $0 < \gamma_U \leq 1$. If x'_{ict} becomes zero, then with a probability p_r the respective asset is replaced with a new asset j which is not yet included (i.e., $b_{jct} = 0$) and is given some random weight $x'_{jct} \in [0, 2 \cdot U_t]$; in addition the respective binary variables are set $b'_{ict} = 0$ and $b'_{jct} = 1$. When selecting j, preferences based on the "Averaged Idol" as introduced later are used. With probability $1 - p_r$ asset i is kept in the portfolio with weight $x'_{ict} = 0$ which means that there are actually less than k assets represented in the portfolio.

Having changed the weights and standardized them such that $\sum_i x'_{ict} = 1$, the fitness of the resulting modified portfolio, $\mathcal{P}'_{ct}$, is calculated. According to the principles of SA, the modifications are accepted with probability $p = \min\left\{1, \exp\left(\Delta F / T_t\right)\right\}$ according to the Metropolis function. $\Delta F = F_{\mathcal{P}_{ct}} - F_{\mathcal{P}'_{ct}}$ is the change in the fitness function (here: objective function) and T_t is the temperature in iteration t. Due to this definition, impairments become less likely with larger decreases in the fitness function and with lower temperatures, i.e., the more iterations have already been passed. The temperature is reduced once per iteration according to $T_t = T_{t-1} \cdot \gamma_T$ where $0 < \gamma_T < 1$ is the cooling parameter. For each crystal, this procedure of generating a modified portfolio and deciding whether to accept it or not is repeated for a fixed number of times.

Evaluation. The crystals are ranked according to their fitness. This evaluation and ranking procedure is crucial for the decision which solutions to reinforce and which portfolios to replace. Akin to the rank based system introduced in Bullnheimer, Hartl, and Strauss (1999) we allow only the best π of all crystals to be "role models" for others (we shall call them *prodigies*). Based on their ranks, the prodigies' portfolios are assigned linearly decreasing *amplifying factors*, a_{ct}, ranging from $\pi + 1$ down to 1. Crystals of the current population that are not prodigies, have an amplifying factor of 0. In addition we enlarge the group of "role models" by ε *elitists* all of which represent the best overall solution found so far.

Replacement. To reinforce promising tendencies on the one hand and eliminate rather disappointing ones on the other, the ω worst crystals of the current population are replaced with crystals that are considered to have promising structures. We distinguish two alternatives of candidates with high potential:

Clone: Based on the amplifying factors, probabilities for selecting an exist-
ing portfolio are calculated such that prodigies with better fitness have a higher probability to be chosen. An unchanged copy of this portfolio replaces the poor portfolio. The effect of cloning is that a new crystal starts off with a supposedly good structure but will ultimately develop a different structure than its twin.

Averaged Idol: An average weight $\bar{x}_{it}$ for each asset is calculated based on the elitists' and prodigies' portfolios. The weights x_{ict} of asset i in prodigy c's portfolio are multiplied by their respective amplifying factors a_{ct}, added to the ε elitists' weights x^*_{it}, and normalized so that the overall sum is 1:

$$\bar{x}_{it} = \frac{\sum_{c \in \Pi_t} a_{ct} \cdot x_{ict} + \varepsilon \cdot x^*_{it}}{\sum_{i=1}^{N} \left(\sum_{c \in \Pi_t} a_{ct} \cdot x_{ict} + \varepsilon \cdot x^*_{it} \right)} \qquad (5.1)$$

where Π_t is the set of the current iteration's prodigies. Usually, the Aver-
aged Idol does not represent a valid solution because more than k assets will have positive weights. These "averaged weights" are therefore used for probabilities to select k assets and assign them weights that again re-
flect these averaged weights (yet with a random component). The effect of this averaging is that an asset is preferred when it is found in many a prodigies' and/or the elitists' portfolio and has a high weight in these portfolios. Unlike in usual Genetic Algorithm systems, there are not just two parents but a whole group of successful ancestors that pass on their endowment.

With a probability of p_c a "Clone" will be chosen, with a probability of $1 - p_c$ a newly generated crystal based on the "Averaged Idol" will be used for a replacement.

The algorithm, summarized in Listing 5.1, stops after a fixed number of iterations and reports the best solution found, i.e., the last elitists' portfolio. The algorithm's runtime is merely influenced by the population size, C, and the maximum number of different assets included in the portfolio, k. The ranking of the crystals can be done with a standard sorting algorithm where the computational complexity is of order

```
FOR c := 1 TO population size DO
    randomly select k assets and assign them positive random weights;
    assign non-chosen assets 0 weight, x_jc := 0.
    standardize weights such that ∑_i x_ic = 1;
end;

FOR t := 1 TO MaxIteration;
    modification (→ Listing 5.2);
    evaluation (→ Listing 5.3);
    replacement (→ Listing 5.4);

    update parameters:
        U_{t+1} := U_t · γ_U ;
        T_{t+1} := T_t · γ_T ;
END;

REPORT elitist;
```

Listing 5.1: Pseudo-code for the main Hybrid Algorithm routine

```
FOR c := 1 TO population size DO
    x'_ic := x_ic;
    FOR changes := 1 TO d DO
        randomly select i with x'_ic > 0;
        find random value for z̃_ic ∈ [−U,+U];
        x'_ic := max{x'_ic + z̃_ic, 0};
        IF (x'_ic = 0) AND (rand < p_r )
            randomly select i with x'_jc = 0;
            find random value for z̃_jc ∈ [0.2 · U];
            x'_ic := x'_jc + z̃_jc;
        END;
    END;
    determine change in the objective function ΔF;
    with probability min{1, exp(ΔF/T)} DO
        x_ic := x'_ic      ∀i;
    END;
END;
```

Listing 5.2: Pseudo-code for the Modification routine of the Hybrid Algorithm

```
rank crystals c according to their current objective value F_c;
determine amplifying factors a_c based on ranks;
check for new elitist (argmax F_c > F_P* ?) ;
```

Listing 5.3: Pseudo-code for the Evaluation routine of the Hybrid Algorithm

```
with each of the w worst crystals DO
    with probability p_c DO
        Clone:
            based on a_c, randomly selected role model c*;
            x_ic := x_ic*;
    otherwise DO
        Averaged Idol:
            replace x_ic with weights according to equation (5.1);
    END;
END;
```

Listing 5.4: Pseudo-code for the Replacement routine of the Hybrid Algorithm

$\mathcal{O}(C \cdot \ln(C))$. For the evaluation of the portfolios, the portfolios' variances have to be calculated. The computational complexity of this is linear in C and quadratic in k, i.e., it comes with a complexity of $\mathcal{O}(C \cdot k^2)$. The algorithm's complexity is at most linear in all other parameters – if affected at all.

5.2.2 Variants

In the course of developing the reported version of the algorithm, we experimented with several modifications and extensions which were eventually turned down for different reasons. Some of these variants are variants to concepts ultimately used, others try to mimic or transfer ideas that have proofed useful in other circumstances.

Variants for the *modification* of portfolios included different ways of changing the weights, x_{ict}, and exchanging assets. Amongst these were alternative versions for calculating $\tilde{z}_{ict}$ which was either distributed within a constant bandwidth (i.e., with constant $U_t = U$) or the distribution of which was distorted towards to the Averaged Idol. Additional versions concentrated on the selection of a new asset j which was either perfectly random or took the covariances σ_{ij} into account. All of these variants led to either lower reliability of the results or increased the runtime significantly without noticeable effect on the quality of results.

Variants for the *evaluation* stage included a nonlinear system for generating the amplifying factors, exclusion of the elitists and/or prodigies, and allowing all members of the current population to contribute to public knowledge and not just the prodigies. Both elitists and prodigies turned out to have positive effects on convergence speed and stability of the results. At the same time, too high a number of agents that contribute their experience to the "Averaged Idol" merely increases runtime without apparent positive effect. For the ranking system, the linear version of the amplifying factors turned out to be both simple and effective.

The *replacement* was found to be most effective when clones of existing good portfolios as well as newly generated portfolios based on successful role models were allowed. To avoid the danger of getting stuck in a local optimum due to "inbreeding" with extremely similar prodigies, we also introduced a third alternative were a random portfolio was generated independently of the other portfolios. Since the other portfolios have already passed a number of steps within the optimization process, this new portfolio was given an extra number of iterations. As turned out, however, this alternative mainly increased the runtime (because of the extra number of iterations) but had hardly any positive effect on the results since these new portfolios almost never made it into the group of the best portfolios with a portfolio structure that differed significantly from some already existing prodigies' portfolios.

5.2.3   Considerations behind the Algorithm

The main goal for the algorithm was to overcome some of the shortfalls of strict local search heuristics such as Simulated Annealing. Though a more or less sophisticated local search heuristic is well capable of finding the proper portfolio weights for assets if there are no "hard" additional constraints, the introduction of cardinality constraints in combination with non-negativity rapidly increases the peril of getting stuck in local optima. In addition, the success of local search methods might well depend on the starting point for the search. Therefore, it appears only natural to overcome these downsides by (i) having more than one starting point and (ii) eliminating solutions that either are already or are likely to get stuck in a local optimum.

This is done by having a population rather than a single agent and by allowing for interaction, role models, and replacement of supposedly poor solutions. Most of the later principles are standard in methods using genetic or evolutionary ideas: Interaction is usually introduced by some form of cross-over, where two parents generate offspring that inherits a combination of (parts of) the parental genetic material. Transfer of good solutions frequently comes together with the extinction of individuals with poor fitness and replacing them with the clone of a better individual.

In the course of experimenting with the different variants we found that for our problem, two significant extensions to the "standard toolbox" of evolutionary principles are helpful: the reinforcement of the elitist concept and the aspect of having role models that goes beyond copying already existing solutions.

The traditional elitist principle ensures that the best solution found so far is maintained. If, in addition, the current best individuals are reinforced as well, the danger of potentially keeping and fostering a local optimum, namely the elitist, is reduced. Also, a strict selection mechanism for eliminating crystals deliberately according to their bad fitness (rather than due to a lower probability for selection as would, e.g., a Genetic Algorithm do), seemed to increase convergence speed. However, whether this later principle is just typical for our optimization problem or a general rule remains to be investigated.

At an early stage the introduction of the "Averaged Idol" allows to find those assets which are most popular in good portfolios found so far. It therefore helps to form the core structure of the portfolio. During the later iterations, it enables the pooling of different good solutions and turned out helpful when the prodigies have found a sound core structure but struggle with the fine tuning, i.e., they have the same assets in their portfolios but with different weights. In both cases, poor crystals can therefore be replaced with new crystals which are not only clones of already existing portfolios but rather of pooled experience.

5.3   The Computational Study

5.3.1   Data and Parameters

The algorithm is tested on the DAX and the FTSE data sets presented in section 4.3.1: Based on daily observations for 30 DAX stocks and 96 FTSE stocks over a period from July 1998 through December 2000, the historic variances, covariances and beta factors were estimated. The volatility of the stocks was set equal to the respective historic volatilities, and the expected return were estimated with the CAPM according to $r_i = r_s + (r_M - r_s) \cdot \beta_i$ with an expected safe return of $r_s = 5\%$, expected market risk premia of $r_{M=DAX} - r_s = 5.5\%$ and $r_{M=FTSE} - r_s = 6\%$, respectively.[11]

To find their appropriate values, we ran experiments with random values for the parameters used in the algorithm for different k's. Since the two data sets differ considerably in their size, we determined separate parameters for the DAX30 and the FTSE100 data sets (the latter reported in brackets when different). The population size is set to 100 (200) where the $\pi = 12$ (15) best portfolios represent the prodigies and where the number of elitists is $\varepsilon = 100$ (200). Each population has 750 (1 000) generations to find the optimum. In each iteration a crystal produces 2 modified portfolios by changing $d = 2$ of the current assets' weights. For the modification of the weights the bandwidth for $\tilde{z}_{ict}$ was $[-U_t; +U_t]$ with $U_0 = 0.3$ and $\gamma_U = 0.9925$. The probability that an asset i with modified weight $x'_{ict} = 0$ was replaced with some new asset was $p_r = 0.4$. In the replacement stage the $\omega = 12$ (15) worst agents were replaced with a clone with probability $p_c = 0.3$ or a new agent based on the "Averaged Idol" with probability $1 - p_c = 0.7$. For the SA part, the initial temperature was $T_0 = 750$, and the cooling parameter was $\gamma_T = 0.97$ (0.95). It is noteworthy, however, that the algorithm appears to be rather "tolerant" with respect to the parameters: exchanging the parameters for the two data sets has no major impact on the quality of the results.

The algorithm was implemented in *Delphi 5*. For the original study, the resulting program was executed on a 900 MHz Pentium III where the runtime was approximately 2 seconds for the smallest problems ($k = 3$) and approximately 45 seconds for the largest ($k = 39$). Executed on a more up-to-date Centrino Pentium M 1.4 GHz, the runtime for the largest problem ($k = 39$) reduces to approximately 14 seconds.

---

[11]   See also sections 3.2.3 and 4.3.1.

5.3.2   Evaluation of the Suggested Algorithm

The graphs in Figure 5.1 depict the results for the case where portfolios consisting of up to $k = 9$ (left column) and $k = 39$ (right column), respectively, out of the $N = 96$ available securities represented in the FTSE100 are to be selected. For the first case there are $\binom{96}{9} = 1.3 \times 10^{12}$ different valid combinations of stocks, in the second case there are $\binom{96}{39} = 1.2 \times 10^{27}$ alternatives. For each of these alternatives, of course, there is an infinite number of different valid combinations of weights since $x_i \in \mathbb{R}_0^+ \ \forall i :$ $b_i = 1$ with $\sum_i b_i = k$. For λ, a random value in the range $[0,1]$ was chosen before each run.

As can be seen from the graphs, the three methods differ in their reliability to find the optimal solutions: Ideally, portfolios should be reported that are on (or at least close to) the so called *efficient set* or *efficient frontier*, the upper border of the area of all theoretically possible portfolios in the volatility-return space. A portfolio is said to be efficient if there is no other portfolio with same constraints that has the same risk but higher expected return or the same expected return but lower risk. Conversely, a portfolio is called *inefficient* or *inferior* if it is below the efficient line. Although the exact efficient line under a given cardinality constraint is not known, a portfolio can definitely be identified to be inferior if it is "south-east" (i.e., right and/or below) of at least one other portfolio with same k in the mean-variance diagram.

The results reported by the method *Simulated Annealing* (SA) (top row in Figure 5.1) form an area rather than a line. This indicates that there are many inferior portfolios that, in addition, are far from optimal. The *GSA* method comes with clear improvements, yet there are still regions where the method apparently reports inferior solutions. Eyeballing suggests that the *Hybrid Algorithm* (HA) is the most reliable of the three methods.

All three algorithms include heuristic local search procedures where the current solutions (i.e., portfolio structures) undergo slight changes. With local search, an asset i will be exchanged for some other asset j by first reducing i's weight, x_i, until i has zero weight and then increasing j's, x_j. This procedure leads to a more or less rapid exclusion of assets with low contribution to the portfolio's fitness. On the other hand, it is less likely that a highly weighted asset i that fits quite good in the current portfolio is replaced with some asset j since the algorithm had to accept a

Method	constraint: $k = 9$	constraint: $k = 39$
SA		
GSA		
HA		

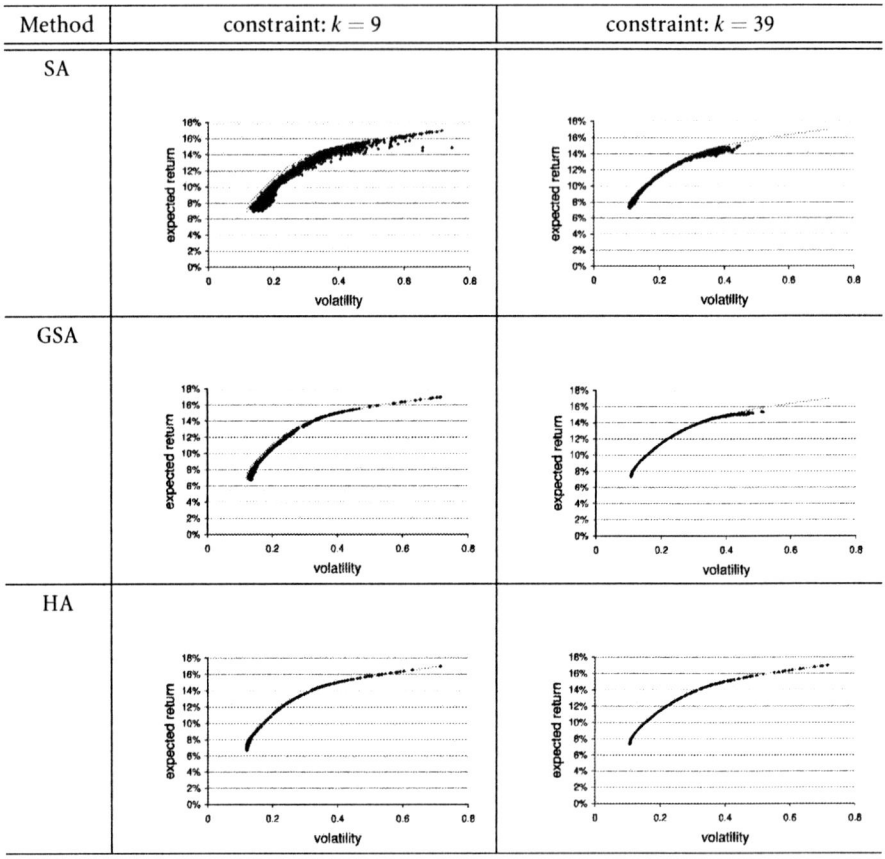

Fig. 5.1: Reported solutions (black) and supposed efficient line (gray) for the FTSE data set depending on different optimization methods and cardinality constraints

series of impairments while eliminating i. Though this allows for a fast evolving core structure, there is also a chance that some asset j which is just slightly better than i remains excluded once the core structure has settled and i is regarded a central part of it. Our heuristic reduces this inherent danger of getting stuck in a local optimum by using a population of crystals rather than isolated agents. Since both assets are equally likely to be selected in the beginning, one can expect that there is the same number of portfolios containing i or j, but those with j have a higher probability to be cloned or enter the Averaged Idol due to their higher fitness. Nonetheless, in some cases the algorithm returns portfolios with optimal weights for a close to optimal selection of assets, i.e., portfolios that are slightly inferior.

As said earlier, the cardinality constraint makes it impossible to determine the exact efficient frontier analytically. One can, however, merge all results from the different methods for the same cardinality constraint k and remove those that can be identified as inferior. The remaining solutions form the sets of best known solutions for the different values of k which can be used as an estimation for the actual efficient line with a cardinality constraint which will be called *supposed efficient line* henceforth.[12]

To assess the dimension of these deviations in particular and the reliability of the algorithm in general, we first determined which of the found solutions are definitively inferior, i.e., which of the portfolios have higher risk with lower expected return than at least one other portfolio. In practice the efficient line can be assumed to be linear between two points given these points are sufficiently close to each other. As the supposed efficient portfolios found by the algorithm form a sufficiently dense line, we determined the two supposed efficient portfolios L and U neighboring an inferior portfolio I, and estimated the deviation Δr_I of the actual return, r_I, from the estimated efficient return $\hat{r}_I^{\text{eff}}$ by linear interpolation:

$$\Delta r_I = r_I - \hat{r}_I^{\text{eff}} = r_I - \left(r_L + \frac{r_U - r_L}{\sigma_U - \sigma_L} \cdot (\sigma_I - \sigma_L) \right)$$

where r_L, r_U, σ_L, and σ_U denote the return and risk of the lower and upper neighbors, respectively. Δr_I therefore indicates, how much more return could be achieved

---

[12] For obvious reasons and as will be discussed in due course, efficient portfolios for a higher k must have at least the same expected return as those with lower k when their volatility is the same. The same is true when the number of assets with non-zero weights rather than the limit k is considered. This property is met by all supposed efficient lines. Nonetheless, being not able to identify a certain portfolio as inferior does – strictly speaking – not imply that it actually is efficient.

	DAX data set			FTSE data set		
k	SA	GSA	HA	SA	GSA	HA
3	−0.952%	−0.174%	−0.001%	−1.722%	−0.631%	−0.064%
6	−0.833%	−0.479%	−0.118%	−1.565%	−0.922%	−0.233%
9	−0.777%	−0.477%	−0.065%	−1.341%	−0.793%	−0.256%
12	−0.682%	−0.417%	−0.127%	−1.179%	−0.646%	−0.189%
15	−0.591%	−0.375%	−0.039%	−1.023%	−0.508%	−0.120%
18	−0.514%	−0.324%	−0.013%	−0.861%	−0.390%	−0.063%
21	−0.455%	−0.278%	−0.001%	−0.738%	−0.321%	−0.046%
24	−0.380%	−0.239%	−0.001%	−0.644%	−0.267%	−0.028%
27	−0.338%	−0.222%	−0.001%	−0.557%	−0.209%	−0.018%
30	—	—	—	−0.474%	−0.150%	−0.009%
33	—	—	—	−0.418%	−0.111%	−0.006%
36	—	—	—	−0.362%	−0.083%	−0.004%
39	—	—	—	−0.320%	−0.072%	−0.005%

Tab. 5.1: Estimated average deviations of inferior portfolios' returns from the respective supposed efficient line in dependence of k

with equal volatility under the same cardinality constraint by choosing a portfolio on the respective supposed efficient line rather than I. Table 5.1 summarizes the estimated average deviations between best known solutions (i.e., the supposed efficient line) and the reported inferior portfolios.

As can be seen from the results, these average deviations are for our suggested algorithm (HA) at most one eighth of an per cent for the DAX data set and at most about a quarter of a per cent for the FTSE data set. It might be surprising, however, that the largest deviations can be found not in the problems with the largest problem space but where $k = 6,...,12$ (DAX data set) and $k = 6,...,15$ assets are to be selected. The main reason for this is the fact that most of the diversification within a portfolio can be achieved with a rather small number of assets. The marginal contribution of any additional asset to the reduction of risk for a given expected return is decreasing. Thus, according to a rule of thumb, a well chosen third of the available assets will result in an efficient line that is already very close to the efficient line of the unconstrained efficient line, whereas portfolios with less assets face increasingly higher risk for a given expected return than the unconstrained efficient line (or an increasing reduction of expected return for a given level of risk). In passing note, that therefore the efficient line without constraints on cardinality and non-negativity cannot be used as a (general) benchmark.

The main problem in finding the optimal solution is to identify the core assets and in due course assigning the proper weights. When k is sufficiently large, chances are that at least some of the populations' crystals find (groups of) these core assets and the applied evolutionary concepts enforce the exchange of information about such groups. A rather small k on the other hand comes with a small number of possible combinations of assets. In both cases, the population quickly identifies the optimal (core) combination and can focus thus on the adjustment of the asset weights. This has positive effects on the quality of the algorithm: not only can most of the optimized portfolios be assumed to lie on or very close to the efficient line, but deviate the apparently inferior ones much less from the best known solutions.

For "medium sized" values of k, more iterations are needed for the selection process in itself leaving less iterations for the fine tuning of the weights. In addition, with this size of the problem space the possibility of getting stuck in a local optimum as described earlier is greatest. Therefore, the portion of inefficient portfolios is highest in this group, too. Nonetheless, the deviations of these inferior portfolios from the supposed efficient line are rather small, especially when compared to the results from other methods, as will be discussed in the next section.

5.3.3   Contribution of Evolutionary Strategies

To determine whether a group of agents outperforms individual agents and whether the use of evolutionary strategies improves the algorithm's performance we implemented the problem with a standard Simulated Annealing (SA) approach where single crystals are to solve the problem. In addition we applied a variant of Simulated Annealing with a group of isolated crystals (GSA) without elimination of unfit individuals and with no ranking system and elitists. GSA is therefore a variant of our algorithm without the stages *evaluation* and *replacement*. Where applicable, SA and GSA used the same values for the parameters as our algorithm. SA had 4 000 runs per k, GSA and our hybrid algorithm (HA) had each 1 000 runs per k.

Table 5.2 compares the portions of inefficient portfolios within the runs for different values of k and data sets. In SA a single agent had 750 (1 000) iterations for

	DAX data set			FTSE data set		
k	SA	GSA	HA	SA	GSA	HA
3	98.75%	73.4%	0.6%	98.9%	85.7%	1.4%
6	100.0%	99.4%	32.9%	99.3%	91.4%	51.6%
9	100.0%	99.8%	41.8%	99.4%	91.9%	57.2%
12	100.0%	100.0%	43.7%	99.7%	91.8%	62.8%
15	100.0%	100.0%	19.0%	99.7%	93.1%	62.9%
18	100.0%	100.0%	27.1%	99.8%	92.6%	56.6%
21	100.0%	99.8%	2.5%	99.9%	95.1%	53.5%
24	100.0%	99.9%	4.0%	99.7%	93.1%	47.8%
27	100.0%	99.9%	2.5%	99.7%	92.7%	47.7%
30	—	—	—	99.7%	92.2%	31.0%
33	—	—	—	99.5%	93.1%	22.0%
36	—	—	—	99.2%	90.0%	17.4%
39	—	—	—	99.0%	91.0%	13.6%

Tab. 5.2: Portion of inefficient portfolios

the DAX (FTSE) data sets, whereas in GSA and HA 100 (200) agents were used per run. Hence it does not surprise that SA finds almost exclusively inefficient portfolios. However, it seems remarkable that GSA is clearly outperformed by the algorithm applying evolutionary principles. E.g., for the DAX data set, the worst case for the HA method is with $k = 12$ where some forty per cent of the portfolios can be considered more or less inferior to the supposed efficient line, but so can all the GSA (and SA) optimized portfolios. For the FTSE data set, virtually all of the SA solutions and at least nine in ten of the GSA solutions are inferior, whereas it is just approximately six in ten or even less, when elimination of weak individuals and orientation to the populations' best individuals takes place.

The boost from evolutionary principles is even more evident when the deviations from the supposed efficient line are analyzed. Considering the case with $k = 6$ for the FTSE data set, it can be seen that on the average a portfolio optimized with HA has a return of 0.120% below the supposed efficient line (0.233% when inferior portfolios only are averaged). With GSA the average deviations are 0.843% (0.922%) when all portfolios (inferior portfolios only) are considered, the respective deviations for SA optimized portfolios are 1.553% (1.565%). Figure 5.2 depicts the average deviations

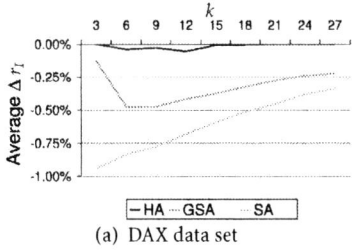
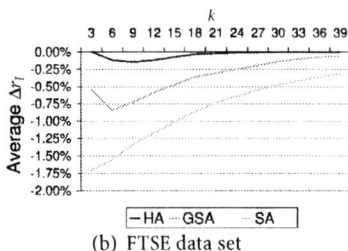

(a) DAX data set   (b) FTSE data set

Fig. 5.2: Average deviation of all optimized portfolios from the respective supposed efficient lines

of all portfolios of the runs per k per method from the supposed efficient line; the results for the average deviation of inferior portfolios were summarized in Table 5.1. The average deviation is the smallest when the portfolios are optimized with HA. Hence, the introduction of evolutionary principles not only reduces the number of portfolios that are clearly inferior but also leads to lower deviations from the supposed efficient line.

5.4 Financial Implications

As mentioned earlier, empirical studies find that investors tend to hold a rather small number of different assets in their portfolios. Figure 5.3 therefore compares the efficient line of the unconstrained problem with the supposed efficient lines of the constrained problems with different values for k. In passing note that, because of the cardinality constraint in combination with the non-negativity constraint on asset weights, the (actual and supposed) efficient lines need no longer to be concave.

For portfolios with high volatility, if existent the differences in the efficient lines are rather small. The reason for this is the non-negativity constraint: in this region there are only few assets an investor should hold with a positive weight, and the cardinality constraint does not come into effect. In low volatility areas, however, strict cardinality constraints come at the cost of reduced returns with equal risk – or having to accept more risk for a given level of expected return. For the FTSE data

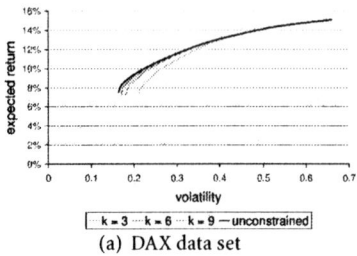

(a) DAX data set

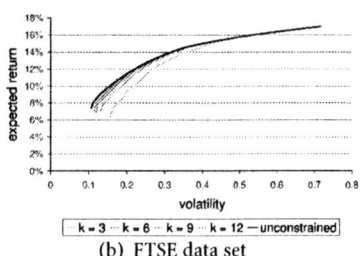

(b) FTSE data set

Fig. 5.3: Supposed efficient lines for constrained portfolios and efficient lines for unconstrained portfolios

set, e.g., the optimal portfolio with a volatility of 0.20 will have an expected return of 9.4% (10.6%; 11.1%; 11.3%) when the cardinality constraint is $k = 3$ (6; 9; 12); without cardinality constraint an investor could expect a return of 11.4%. On the other hand, the optimal portfolios for an expected return of 8% will have volatilities of 0.177 (0.141; 0.130; 0.122) for $k = 3$ (6; 9; 12) whereas in the unconstrained case the optimal solution would be 0.107.

For the FTSE data set, the differences between unconstrained portfolios and constrained portfolios with $k > 15$ are virtually negligible, for the DAX data set, optimized portfolios with $k > 12$ are virtually as good as the unconstrained – provided that the portfolio structure is determined with a reliable optimization method. Simply speaking, reasonable diversification is possible with few, yet well-chosen assets. In a world where transactions costs, information gathering costs and portfolio management fees depend on the number of different included assets, a cardinality constraint might be crucial for successful investment.

The financial implications from this empirical study therefore confirm those found in the previous chapter where the case of identifying the constrained portfolio with the maximum Sharpe Ratio is considered. The optimal portfolios identified in the previous chapter are particular members of the respective Markowitz efficient sets with the same cardinality constraint that have been identified in this chapter.

5.5   Conclusion

In this chapter a meta-heuristic was presented that basically combines principles from Simulated Annealing with evolutionary strategies and that uses additional modifications. Having applied this algorithm to the problem of portfolio selection when there are constraints on the number of different assets in the portfolio and non-negativity of the asset weights, we find this algorithm highly efficient and reliable. Furthermore, it is shown that the introduction of evolutionary principles has significant advantages.

The algorithm is flexible enough to allow for extensions in the optimization model by introducing additional constraints such as transaction costs, taxes, upper and/or lower limits for weights, alternative risk measures and distributions of returns, etc. First tests with such extensions led to promising results and supported the findings for the algorithm presented in this chapter.

Chapter 6

The Hidden Risk of Value at Risk

6.1 Introduction

When it comes to measuring the risk of financial assets, volatility, i.e., the standard deviation of the returns, is a common choice. Meanwhile, this concept of risk has been adopted by almost all participants in the investment industry and is the foundation for many a seminal piece of academic work with the foundations being formalized in the 1950's by H. Markowitz in his Modern Portfolio Theory and its assumption of normally distributed returns. However, the finance and financial econometrics literature raised serious doubts that this assumption holds: hardly any time series of returns can be described reliably with mean and variance only, and the existence of skewness, excess kurtosis, and autocorrelation seems to be the rule rather than the exception.

The decision whether to rely on the normality assumption or not becomes even more important with the introduction of alternative risk measures such as *Value at Risk* (*VaR*, also known as *Capital at Risk* or *Money at Risk*)[1], which describes the loss that is not exceeded with a given probability α, or *Expected Tail Loss* (*ETL*, also known as *Conditional Value at Risk*, *Mean Excess Loss*, or *Expected Shortfall*), which is to measure the expected loss given that one encounters a loss at or beyond the VaR

---

[1] See Riskmetrics Group (1996). Jorion (2000) and Simons (2000) present concepts and applications. Longin (2000) compares VaR to *stress testing* which is also concerned with rare yet hazardous situations where extreme losses may occur. In this respect, see also Keenan and Snow (2002) and Gilli and Këllezi (2003).

threshold.[2] The VaR risk measure gained additional relevance as the Basel accord accepts it as a valid risk measure for estimating banks' exposure to credit risk.[3]

Much akin to the traditional mean-variance framework, the objective for portfolio optimization with alternative risk measures can either be the minimization of risk given a constraint on the expected return or wealth (if considered at all)[4] or the maximization of the expected (utility of the) return (wealth) with a constraint on the risk measure.[5]

To estimate the VaR, three basic methods are employed: *parametric estimation* assumes that the assets' returns follow a parametric distribution with known parameters; *historic simulation* takes past realizations and assumes that their empirical distribution is apt to describe future outcomes; and *Monte Carlo simulation* generates prices based on a parametric and/or empirical distribution. The use of empirical distributions or Monte Carlo approaches leads to apparently better results than the assumption of normally distributed returns[6] – yet at the cost of considerably higher computational complexity. This is all the more true for models that use VaR rather than ETL. The latter has much more pleasant properties[7] whereas the first one usually comes with a large number of local optima and is difficult to solve due to its non-convexity: As shown by Daníelsson, Jorgensen, de Vries, and Yang (2001), the portfolio optimization problem under a general VaR constraint is NP hard. Hence,

---

[2] For a discussion of these concepts and related approaches, see Manganelli and Engle (2001). Frey and McNeil (2002) address these risk measures with respect to credit risk.

[3] See Basel Committee on Banking Supervision (2003, §§109, 149—152 and 490). For a discussion of related literature, see Alexander and Baptista (2001). Berkowitz and O'Brien (2002) empirically test the accuracy of VaR models in commercial banks; their "findings indicate that banks' 99th percentile VaR forecasts tend to be conservative, and, for some banks, are highly inaccurate" (p. 1108).

[4] See, e.g., Rockafellar and Uryasev (2000) (extended in Krokhmal, Palmquist, and Uryasev (2001)), Uryasev (2000), or Pflug (2000).

[5] See, e.g., Arzac and Bawa (1977) (following Roy (1952)), or more recently Campbell, Huisman, and Koedijk (2001) (and the comment by Huang (2004)), and Basak and Shapiro (2001).

[6] See, e.g., Pritsker (1997) or Lucas and Klaasen (1998).

[7] See Pflug (2000) and Rockafellar and Uryasev (2002). For a critical discussion of VaR see Daníelsson, Embrechts, Goodhart, Keating, Muennich, Renault, and Shin (2001).

the literature so far tends to confine themselves to a rather small set of assets included in the portfolio when using empirical distributions.[8]

An alternative way out of this dilemma might be the use of heuristic optimization methods. Dueck and Winker (1992) were the first to solve portfolio choice problems with a heuristic method, namely Threshold Accepting[9] (TA). Gilli and Këllezi (2002) build on this approach and with the same heuristic tackle a portfolio optimization problem where the expected wealth is to be maximized with constraints on the number of assets included, with lower and upper limits to the included assets' weights, demanding the number of assets to be integers – and constrain the shortfall probability for a given Expected Shortfall and VaR, respectively. Their results show that this optimization problem can be handled with heuristic methods. Maringer and Winker (2003) approach a similar optimization problem with a modified version of Memetic Algorithms (MA), a search heuristic combining local and global search methods which we will follow in this article.

Based on the findings in Maringer and Winker (2003), Maringer (2003a, 2005), and Winker and Maringer (2003), the aim of this chapter is to study some of the consequences that arise from the assumed distribution of the asset returns and the choice of the risk constraint. The following section presents a formalization of the optimization problems and a description of the data used for the empirical studies. In section 6.3, the modified version of Memetic Algorithms, applied for this optimization problem, is presented. In section 6.4 a computational study for the stock market data set is presented where the results due to different risk constraints are compared, section 6.5 reports a corresponding empirical study for a bond market. The chapter concludes with a short summary of the main results and their consequences.

---

[8] See also Andersson, Mausser, Rosen, and Uryasev (2001) and Krokhmal, Palmquist, and Uryasev (2001) who suggest the use of linear programming for simplified frameworks under ETL and VaR, respectively.

[9] See Dueck and Scheuer (1990), Winker (2001), and section 2.3.1.

6.2 Risk Constraints and Distribution Assumptions

6.2.1 The Stock Market Investor

6.2.1.1 The Optimization Model

Following standard assumptions about behavior of a risk averse and rational investor making myopic investment decisions, the objective for the stock market participant is to maximize the expected utility of the investment's return, r, (which, in particular for a unit investor, is equivalent to the (logarithmic) utility of the portfolio's wealth) in one period of time without possibilities for restructuring the portfolio in between. The investor's decision variables are mainly the assets' weights within the portfolio which must not be negative and sum up to 1. In addition, the investor can rely on the validity of the separation theorem[10] which, for the current problem, allows to split her endowment and to invest a fraction q into a risky portfolio $\mathcal{P}$ with return $r_{\mathcal{P}}$ and volatility $\sigma_{\mathcal{P}}$ and the remainder of $(1 - q)$ into a risk-free asset which has a safe return of r_s. The fraction q is chosen in a way the risk constraint is met.

Depending on the risk constraint (*VaR* and *ETL* for Value at Risk and Expected Tail Loss, respectively) and the assumed distribution of the assets' returns (with superscripts *emp* and *norm* for empirical and normal, respectively), the investor's optimization problem can therefore be summarized as follows:

$$\max_{x_i} E(\mathcal{U}(r))$$

subject to

$$r = q \cdot r_{\mathcal{P}} + (1 - q) \cdot r_s$$

$$r_{\mathcal{P}} = \sum_i x_i \cdot r_i$$

$$\sigma_{\mathcal{P}} = \sqrt{\frac{T}{T-1}\left(E(r_{\mathcal{P}}^2) - (E(r_{\mathcal{P}}))^2\right)}$$

$$\sum_{i=1}^{N} x_i = 1$$

---

[10] See, e.g., De Giorgi (2002).

$$x_i \geq 0 \quad i = 1...N$$

and one of the risk constraints

$$VaR^{emp} : prob\,(r \leq r_{VaR}) = \alpha$$

$$\Longleftrightarrow q \cdot r_{\tau P} + (1-q) \cdot r_s = r_{VaR} \text{ where } \tau = \alpha \cdot T$$

$$ETL^{emp} : E\,(r|r \leq r_{VaR}) = r_{ETL}$$

$$\Longleftrightarrow q \cdot \left(\frac{1}{\alpha \cdot T} \sum_{\tau=1}^{\alpha \cdot T} r_{\tau P} \right) + (1-q) \cdot r_s = r_{ETL}$$

$$VaR^{norm} : prob\,(r \leq r_{VaR}) = \alpha$$

$$\Longleftrightarrow q \cdot (E(r_P) - |u_\alpha| \cdot \sigma_P) + (1-q) \cdot r_s = r_{VaR}$$

$$ETL^{norm} : E\,(r|r \leq r_{VaR}) = r_{ETL}$$

$$\Longleftrightarrow q \cdot \left(E(r_P) - \frac{\phi(u_\alpha)}{\alpha} \cdot \sigma_P \right) + (1-q) \cdot r_s = r_{ETL}$$

where r is the (overall) return from the investment into safe asset and risky portfolio. $r_{\tau P}$ denotes the τ-th worst return of the T observed days. $\mathcal{U}(\cdot)$ is the (logarithmic) utility function, u_α is the α-quantile of the normal distribution such that $\Phi(u_\alpha) = \alpha$, and $\phi(\cdot)$ returns the density of the normal distribution. In this model, r_{VaR} is fixed *a priori*, hence q can be determined by rearranging the risk constraints.

6.2.1.2 The Stock Market Data

The data base for the computational study consists of the daily returns of the Standard and Poor's 100 stocks over the period from November 1995 through November 2000 which is known to cover a time span with a rather particular market situation. In order to find a suitable trade-off between reasonable length of in sample periods and the peril of over-estimation, the in sample period is set to 200 trading days and the number of risky assets in the portfolio $\mathcal{P}$ is 25. We generated a number of cases by randomly picking a starting point for the time frame and then randomly selecting the assets. For each case, the out of sample periods were 1, 10, 20, 50, 100, and 200 trading days following the in sample period. While the optimization method is well capable of dealing with portfolios that have more assets and longer in sample observations, such portfolios might have lead to undesired problems: more assets would demand for more observations (i.e., longer in sample periods in the lack of high frequency observations) to avoid linear dependencies in the (by definition) small number of days where the VaR limit is exceeded; longer observation periods, on the other hand, are likely to cause additional specification errors as both the parametric and the empirical distributions exhibit instabilities in the considered time frame.

If not stated otherwise, r_{VaR} is set to -0.005 per day for the VaR models; for the ETL models, the expected loss on the $(\alpha \cdot T)$ worst of the T trading days shall be $r_{ETL} = -0.005$. The safe return, r_s, is the 3 month EURIBOR from the day follow-ing the case's last in-sample day (i.e., the first day of the out of sample period).[11] We allow a shortfall probability of $\alpha = 0.10$; under the normality assumption, the respec-tive parameters are therefore $|u_\alpha| = 1.2816$ and $\phi(u_\alpha)/\alpha = 1.7550$, and for empirical distributions, the "worst" 20 out of the 200 in sample trading days are considered. In real life problems, α is usually noticeably lower. In particular when empirical distri-butions are used, this causes additional problems as it demands more observations: To have τ observations to represent the scenarios with returns at or below the VaR limit, the in sample data set needs to consist of τ/α observations. Hence, (as in our case) $\tau = 20$ with a shortfall probability of $\alpha = 1\%$ (or 0.5%) would require at least 2 000 (4 000) in sample observations per included asset; for portfolios with more as-sets than in our case, higher values for τ are necessary for reliable results which again increases the required overall number of observations per asset. As for the majority of assets there are not enough real and reliable data available,[12] the "real" data are often complemented with artificially generated scenarios coming from some data generating process – which, however, demands additional assumptions that might cause additional problems.

The main computational study includes 250 portfolios generated as described above. Each of these portfolios was optimized under the different risk constraints and with the presented version of the MA. To minimize the risk of reporting lo-cal optima, each problem had repeated and independent runs, the best solutions of which are used for the following analyses.

---

[11] An alternative would be to use risk premia instead of returns in order to account for changes in the risk free rate. In preliminary tests, however, this did not improve the results.

[12] Note that the data set ought to contain neither too old observations nor too dense data. The former are not necessarily valid representatives for the immediate VaR horizon. The latter refers in particu-lar to high frequency data which are observed in time intervals that are significantly smaller than the VaR horizon. These data often exhibit particular additional properties and are therefore not always representative either.

6.2.2 The Bond Market Investor

6.2.2.1 The Optimization Model

Unlike for stocks, bonds are usually traded at lot sizes, and investors might therefore be more restricted in choosing the portfolio weights. The investor for our problem has an initial endowment of V_0 that can be either invested in bonds or kept as cash; without loss of generality, the rate of return of the latter is assumed to be zero. Given that the losses until time τ must not exceed a (fixed) value of $\underline{\delta}^{VaR} \cdot V_0$ with a given probability of α, and that there are no other risk constraints, a manager of a bond portfolio will be inclined to find a combination that has maximum expected yield that does not violate this VaR constraint.

The optimization model can therefore be written as

$$\max_{n_i} E\left(r_P\right) = \sum_i \frac{n_i \cdot L_i \cdot D_{i,0}}{V_0} \cdot r_i$$

subject to

$$n_i \in \mathbb{N}_0^+ \quad \forall i$$
$$\sum_i n_i \cdot L_i \cdot D_{i,0} \leq V_0$$
$$prob\left(V_\tau \leq V_0 \cdot \left(1 - \underline{\delta}^{VaR}\right)\right) = \alpha$$

where L_i and $D_{i,0}$ are lot size (in monetary units) and current clean price (in per cent), respectively, of bond i, and r_i is its yield to maturity *per annum*. n_i is the number of lots kept in the portfolio which has to be non-negative; also, the cash position must be non-negative. V_τ is the value of the portfolio at time τ (i.e., the value of the bonds including accrued interest from time 0 to τ) plus cash.

For estimating V_τ, we apply the following methods:

- Assuming *normal distribution*, the VaR constraint can be rewritten as

$$E\left(V_\tau\right) - u_\alpha \cdot \sigma_{V_\tau} \geq V_0 \cdot \left(1 - \underline{\delta}^{VaR}\right)$$

  where u_α is the respective quantile of the standard normal distribution. The expected value for V_τ and its volatility are alternatively estimated from past observations either in a standard way ("plain vanilla" or "pv" henceforth) or with weighted values where more recent observations contribute stronger. The latter version turned out advantageous for stock portfolios in a similar setting[13] with a decay factor of 0.99 which is applied here, too. The weights are therefore $w_s = \frac{0.99^{(S+1)-s}}{\sum_{t=1}^{S} 0.99^t}$ where the simulations are ordered chronologically and $s = 1$ is the simulation based on the oldest, $s = S$ on the most recent of the S observations.

- Assuming *empirical distribution*, the VaR constraint can be rewritten as

$$\sum_{s=1}^{S} b_s \leq \alpha \quad \text{with } b_s = \begin{cases} \frac{1}{S} & \text{if } V_s \leq V_0 \cdot \left(1 - \underline{\delta}^{VaR}\right) \\ 0 & \text{otherwise} \end{cases}$$

where $V_{s,\tau}$ is one out of S simulations for the wealth at time τ based on historic (in sample) observations. To parallel the weighted version of the normal distribution, the b_s's can be computed in a way to reflect the "timeliness" of the observations:

$$\sum_{s=1}^{S} b_s \leq \alpha \quad \text{with } b_s = \begin{cases} w_s = \frac{0.99^{(S+1)-s}}{\sum_{t=1}^{S} 0.99^t} & \text{if } V_s \leq V_0 \cdot \left(1 - \underline{\delta}^{VaR}\right) \\ 0 & \text{otherwise} \end{cases}$$

where, again, simulation $s = S$ is based on the most recent, $s = 1$ on the oldest observation.

For the main computational study presented in the following sections, the investor will be endowed with $V_0 = $ CHF 1 000 000, and the VaR constraint demands that the next day's wealth will not be below 990 000 (i.e., $\underline{\delta}^{VaR} = 0.01$) with a probability of $\alpha = [0.025; 0.05; 0.1]$. In addition, alternative values for these parameters were employed; as they merely confirmed the qualitative results reported for the main study, they are omitted in the sense of brevity.

6.2.2.2 The Bond Market Data

The computational study for the bond market investor is based on bonds with fixed coupon quoted on the Swiss stock exchange in local currency, i.e. CHF. From all

---

[13] See Maringer (2003b) and section 6.4.

quoted bonds, we chose randomly 42 Swiss and 113 foreign issuers, yet it was sought that no industry sector or issued volume is over- or under represented. For these bonds, we have daily (clean) closing prices (when traded) for the period January 1999 through June 2003. In particular for the earlier part of this time series, thin trading causes many missing data – the respective yields to maturity were eventually estimated based on the correct time to maturity and by using the previously quoted price.

From this data set, random selections of bonds were drawn by first choosing a random date and then selecting $N = 10$ (20) different bonds. Any of these selections was accepted only if a minimum number of different quotes within the in sample as well as the out of sample time frame were observed (in sample frame: chosen date plus 200 in sample days; out of sample frame: the subsequent 100 trading days). For both values of N, 250 of such case sets were generated independently.

6.3   A Modified Version of Memetic Algorithms

6.3.1   Memetic Algorithms and Threshold Accepting

Memetic Algorithms (MA)[14] are an evolutionary population based search heuristic where the agents *search locally* (i.e., modify their current solution), *cooperate* with some agents (i.e., produce new solutions by means of a cross-over operator) and *compete* with other agents (i.e., challenge their one immediate neighbor and are challenged by their other immediate neighbor).[15] Compared to preceding, "traditional" evolutionary concepts, the main gist of MA is not only the stronger emphasis on local search, but also that any decision over whether an individual's current structure is replaced with a new structure (be it the modified one, be it the challenger's) is based on the same decision function used for the local search part. Compared to "traditional" local search strategies, the "interaction steps" of the heuristic reduce the risk of getting stuck in local optima or having a solitaire agent getting lost in

---

[14] See Moscato (1989) and Moscato (1999).

[15] See also the presentation in section 2.3.4.

the solution space. The modifications suggested in this chapter are to use a different local search principle and to reinforce the current optimum.

In MA, a common method of choice for local search is *Simulated Annealing* (SA)[16] where the probability for accepting a modification comes from the Metropolis function

$$p = \min\left(1, \exp\left(\Delta F / T_i\right)\right)$$

where ΔF is the difference between the current and the new value of the objective function F of an maximization problem, and T_i is the temperature in the iteration i which is lowered with a cooling factor γ according to $T_i = T_{i-1} \cdot \gamma$. With the Metropolis function, improvements are always accepted and impairments are rejected randomly (see also Figure 6.1(a)).

Alternatively, the Boltzmann function

$$p = \frac{1}{1 + \exp\left(-\Delta F / T_i\right)}$$

can be used. Here, improvements are likely to be accepted ($0.5 < p < 1$) and impairments are likely to be rejected ($0 < p < 0.5$), yet in either case the decision is stochastic (see Figure 6.1(b)).

A variant of Simulated Annealing is *Threshold Accepting* (TA)[17] where improvements are accepted at any rate – and so are impairments given they do not exceed a certain threshold. If the impairment exceeds the threshold, the change is rejected. TA therefore has a deterministic acceptance criterion. In the course of iterations, the threshold is consecutively lowered making the acceptance function less tolerant to impairments in the objective function (see Figure 6.1(c)).

While the asymptotic convergence results for Threshold Accepting are similar to those for the Simulated Annealing algorithm,[18] the few existing comparative implementations seem to indicate a slight advantage for Threshold Accepting.[19] A possible disadvantage of TA is the fact that no standard sequence of threshold values is

---

[16] See Kirkpatrick, Gelatt, and Vecchi (1983).

[17] See Dueck and Scheuer (1990) and Winker (2001).

[18] See Althöfer and Koschnik (1991).

[19] See Winker (2001, pp. 109ff).

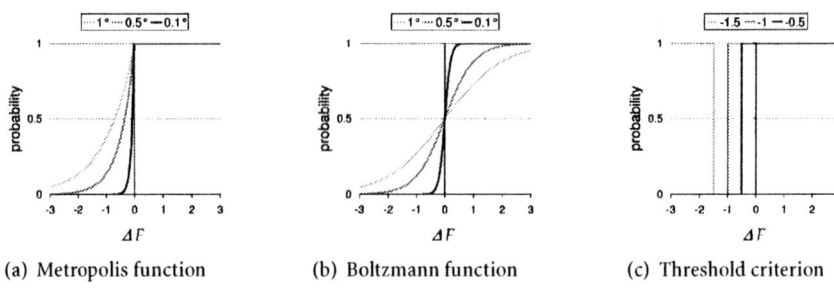

(a) Metropolis function   (b) Boltzmann function   (c) Threshold criterion

Fig. 6.1: Acceptance probabilities for criteria depending on the change in a maximization problem's fitness function, ΔF, for different "temperatures" and thresholds

available comparable to the geometric cooling schedule for SA. However, given that the thresholds have an intuitive interpretation as local differences of the objective function, a threshold sequence can be constructed from a simulated distribution of such local differences (see Winker (2001, pp. 145f), for details). This approach is also followed for the present implementation.

6.3.2   The Elitist Principle

Both SA and TA apply acceptance criteria that allow moving away from the optimum at the risk of leaving it for good and "not finding back". This is acceptable or even advantageous when the optimum is a local one which is to be left, yet baffling when the agent is already close to the global optimum or when the problem space is rather smooth. It might therefore be desirable to remember what has been found to be the optimum so far as this allows determining whether the current solution is obviously a local one and inferior to already found solutions.

To incorporate a reminder of that past success, we suggest considering not only the agents that make up the current population, but also one additional individual, that is neither part of this population nor behaving like all other individuals. This individual represents the best solution found so far and, in line with the naming conventions in other methods,[20] is denoted *elitist*. In order to keep the main MA

---

[20] See, e.g., Dorigo, Maniezzo, and Colorni (1996) and section 4.2.2.

framework unchanged, we allow the elitist to appear only at the competition stage where she challenges one agent from the population. Whether this agent will or will not take the elitist's solution is decided with the same probabilistic principle as for the "standard" challenges.

By definition, the elitist is at least as good as any of the individuals from the current population. Under SA's Metropolis function as well as under TA, being challenged by the elitist means certain replacement of the challenged individual's current solution. Hence, either approach enforces a hill-climbing strategy as sooner or later every individual will be set back to the current elitist's structure. This might be advantageous and increase the convergence speed if the problem space is rather smooth without local optima. However, the more local optima there are, the greater the chances that the elitist represents a local optimum; and reinforcing a local optimum makes it difficult for the individuals to escape it. Remedies for this downside are a larger population size (which also prevents a meme representing some local optimum and passing it quickly on to the whole population) or having the elitist challenge more seldom.[21]

A central question is how to select the agent that is challenged by the elitist. One alternative would be to virtually place the elitist on the ring together with the actual individuals, but where she becomes "visible" only when it is her turn to challenge and ignored in all other situations. This would imply that the elitist always challenges the same agent who, whatever the probability function, has a higher probability of accepting rather than rejecting the elitist's structure. In the long run this might reinforce the caveats discussed previously for the acceptance functions. The alternative would therefore be that the elitist keeps on challenging different agents, selecting them either according to some deterministic rule or randomly. In our implementation, we went for the perfectly random selection as this performed quite well in preliminary tests and consumed the least computing time.

---

[21] As indicated, the elitist will impose her structure on the challenged agent under both SA's Metropolis function and TA. One way to avoid this sure replacement would be the use of a probability function where improvements usually have a high, but not a hundred per cent probability to win the conquest – as is the case with the Boltzmann function. One might expect that this function lowers the peril of getting stuck in a local optimum as challenged agents with good structure have some chance to not be imposed with the elitist's. In preliminary tests, however, we could not find a statistically significant advantage of one function over the other – and therefore stick to the original Metropolis function for SA and a negative threshold for TA, respectively.

```
FOR m := 1 TO Population size;
    x_m := ValidRandomStructure;
END;
initialize parameters and variables including acceptance criteria;

FOR i := 1 TO MaxIterations
    perform neighborhood search (→ Listing 6.3);
    compete (→ Listing 6.4);
    perform neighborhood search (→ Listing 6.3);
    cooperate (→ Listing 6.5);
    adjust Acceptance criterion (SA: Temperature; TA: Threshold);
END;
```

Listing 6.1: Pseudo-code for the modified Memetic Algorithm

```
FUNCTION AcceptOver(x^new, x^current): Boolean;
    determine difference in fitnesses:
        ΔF = F(x^new) − F(x^current);

    return TRUE when the following is true (depending on criterion):
        TA:                      ΔF > Threshold
        SA, Metropolis function:  rand < exp(ΔF/T_i)
        SA, Boltzmann function:   rand < (1+exp(−ΔF/T_i))^−1
    return FALSE otherwise;
```

$$\Delta F = F(x^{new}) - F(x^{current});$$

$$TA: \quad \Delta F > \text{Threshold}$$

$$SA, \text{Metropolis function:} \quad \text{rand} < \exp\left(\Delta F / T_i\right)$$

$$SA, \text{Boltzmann function:} \quad \text{rand} < \left(1 + \exp\left(-\Delta F / T_i\right)\right)^{-1}$$

Listing 6.2: Pseudo-code for the acceptance criterion

```
FOR m := 1 TO Population size;
    x'_m := x_m;
    randomly select two assets i and j;
    without violating the constraints
        lower x'_im by random amount;
        increase x'_jm correspondingly;

    IF AcceptOver(x'_m, x_m) = true (→ Listing (6.2)) THEN
        x_m := x'_m;
    END;
END;

Check whether new elitist has been found;
```

Listing 6.3: Pseudo-code for the neighborhood search procedure in the Modified Memetic Algorithm

```
for later use, remember meme m = 1:
    xₕ := x₁;

FOR m := 2 TO Population size;
    IF AcceptOver(xₘ, xₘ₋₁) = true (→ Listing (6.2)) THEN
        xₘ₋₁ := xₘ;
    END;
END;


compare first (before replacement) and last meme:
m := Population size;
IF AcceptOver(xₘ, xₕ) = true (→ Listing (6.2)) THEN
    xₘ := xₕ;
END;


elitist principle, if applicable:
with probability pₑ
    randomly select one meme, m;
    IF AcceptOver(xᵉˡⁱᵗⁱˢᵗ, xₘ) = true (→ Listing (6.2)) THEN
        xₘ := xᵉˡⁱᵗⁱˢᵗ;
    END;
END;
```

Listing 6.4: Pseudo-code for the competition procedure in the Modified Memetic Algorithm

Listing 6.1 summarizes the main steps of the algorithm. The algorithm's computational complexity is merely determined by the risk constraint which has to be checked for each candidate solution. Under empirical distributions, the S in sample observations portfolio returns have to be determined (complexity: $\mathcal{O}(k \cdot S)$) and sorted (complexity: $\mathcal{O}(S \cdot \ln(S))$) to find the worst days. Under the normal distribution, computing the portfolio's variance via the assets' covariance matrix, it demands $\mathcal{O}(N^2)$, when determined by first computing the portfolio's daily in sample returns and then finding their variance, the complexity is $\mathcal{O}(S \cdot (k + 4))$. Unlike in Genetic Algorithms, the individuals of a MA need not be ranked, thus the complexity is linear in the population size and, as usual, linear in the number of iterations.

In a computational study, we tested the SA and TA acceptance functions with and without the elitist principle. The results are presented in the next section.

```
HalfPopSize := Population size / 2
FOR m := 1 TO HalfPopSize;
    generate vector with N random (binary/equally distributed) variables bᵢ,
        b := [bᵢ]₁ₓₙ;
    mother = m;
    father := (m + HalfPopSize);
    FOR i := 1 TO N DO
```

$$x_i^{daughter} := b_i \cdot x_i^{mother} + (1 - b_i) \cdot x_i^{father} \;;$$
$$x_i^{son} := (1 - b_i) \cdot x_i^{mother} + b_i \cdot x_i^{father} \;;$$

```
    END;


    Ascertain that new weights are valid w.r.t. constraints:
```
$$x^{daughter} := \texttt{Valid}(x^{daughter});$$
$$x^{son} := \texttt{Valid}(x^{son});$$

```
    IF AcceptOver(xᵈᵃᵘᵍʰᵗᵉʳ, xᵐᵒᵗʰᵉʳ) = true THEN
```
$$x^{mother} := x^{daughter}; \quad \texttt{END};$$

```
    IF AcceptOver(xˢᵒⁿ, xᶠᵃᵗʰᵉʳ) = true THEN
```
$$x^{father} := x^{son}; \quad \texttt{END};$$

```
END;


check whether new elitist has been found;
```

Listing 6.5: Pseudo-code for the cooperation procedure in the Modified Memetic Algorithm

6.3.3 Computational Study

For the initial test-bed for the MA, we randomly selected 10 stock portfolios as described in section 6.2.1.2. Each of these portfolios had to be optimized under the VaR^{emp}, ETL^{emp} and the VaR^{norm} risk constraint. For obvious reasons, the VaR constraint for empirically distributed returns has the roughest of the solution surfaces. ETL has fewer local optima, and the assumption of normally distributed returns additionally smoothens the solution landscape. Each of the optimization problems was solved in about 200 independent runs per variant of the algorithm and per risk constraint. As a version is *a priori* more desirable when it identifies solutions with higher values of the objective function, we compared pairs of versions and tested whether the differences in the reported solutions are statistically significant.[22]

---

[22] For a general discussion and methodological aspects for the comparison of different heuristics, see, e.g., Barr, Golden, Kelly, Resende, and Stewart, Jr. (1995).

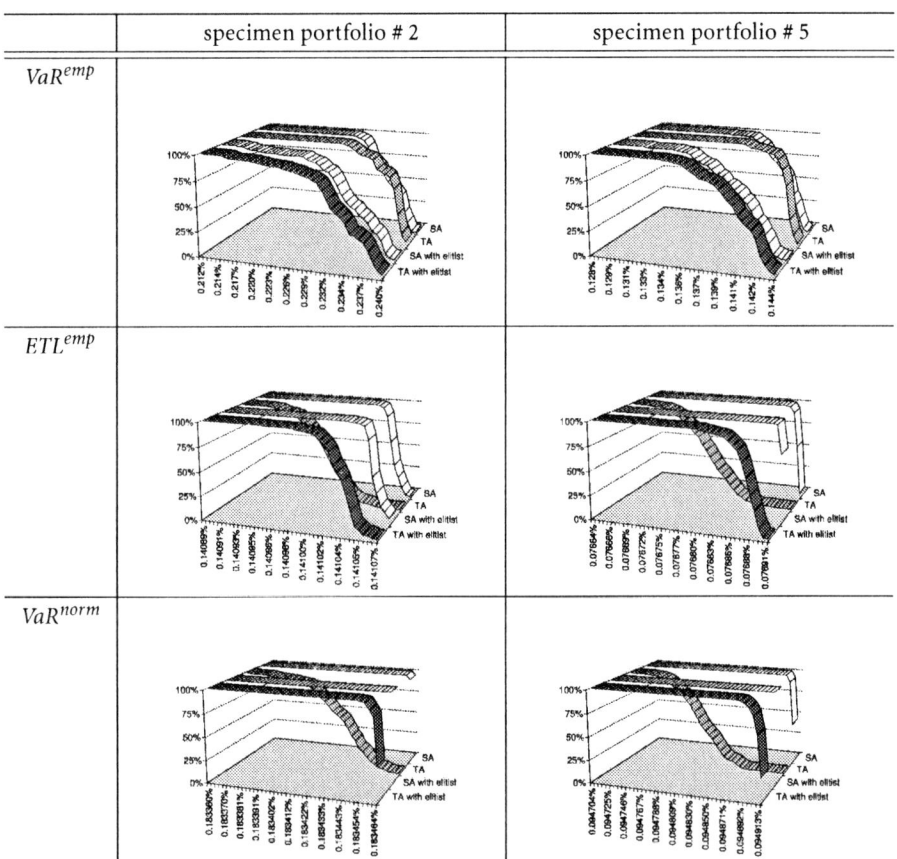

Fig. 6.2: Cumulated share of reported solutions reaching at least a certain value for the objective function for different risk constraints for two of the ten specimen portfolios

constraint		SA	SA, elitist	TA, elitist
VaR^{emp}	TA	TA (0.01%)	TA (0.00%)	TA (0.00%)
	TA, elitist	SA (0.86%)	TA, el. (20.42%)	—
	SA, elitist	SA (1.36%)	—	—
ETL^{emp}	TA	SA (0.00%)	SA, el. (0.00%)	TA, el. (0.00%)
	TA, elitist	SA (0.00%)	SA, el. (0.00%)	—
	SA, elitist	SA, el. (10.16%)	—	—
VaR^{norm}	TA	SA (0.00%)	SA, el. (0.00%)	TA, el. (0.00%)
	TA, elitist	SA (3.36%)	SA, el. (0.00%)	—
	SA, elitist	SA, el. (0.00%)	—	—

	rank			
constraint	1	2	3	4
VaR^{emp}	TA	SA	TA, el.	SA, el.
ETL^{emp}	SA, el.	SA	TA, el.	TA
VaR^{norm}	SA, el.	SA	TA, el.	TA

Tab. 6.1: "Winner" of the pair-wise comparisons of SA and TA without and with elitist strategy, respectively (brackets: average p values for H_0: average fitness is equal for both versions vs. H_1: different means, from a t-test allowing for heteroscedasticity in the samples)

As can be seen from the statistics in Table 6.1 and the specimen cases presented in Figure 6.2, the computationally most demanding risk constraint, VaR^{emp}, comes with the broadest bandwidth of reported solutions. Among the four tested versions for the MA, using the Threshold Accepting principle generally brings the best result in the sense that on average the reported optima are better than those reported from the other versions, and that they are rather close together. This implies that one can expect results produced with the TA version to be at or very close to the (supposed) global optimum.[23] Using SA's Metropolis function for local search and the acceptance decisions yields almost equally good results, though with a slightly higher chance of being off-optimal. For either acceptance criterion, the introduction of the elitist principle tends to reduce stability. A closer look at these portfolios' structures confirms that the elitist might increase the chances of getting stuck

---

[23] As we do not have a reference problem with known exact solution, we cannot be sure whether the best found solution in any of the runs actually is the global optimum or not.

in a local optimum if this solution has an objective value only slightly below the (supposed) global optimum yet with a rather different portfolio structure. In this case, leaving the local optimum and moving towards the global one would demand a series of accepted impairments. Yet, moving far from any optimum increases the chances of being replaced with the elitist's structure when challenged.

ETL has not only more desirable properties from a theoretical point of view,[24] it also has a less demanding problem space than VaR when the empirical distribution is used. Hence, the bandwidth of the reported solutions narrows as heuristic methods have a better chance of finding the optimum: whereas for VaR^{emp}, the average ratio between the optimum and median of the reported solution is 1.105, it is just 1.001 for the ETL^{emp} problems. Comparing the different versions of MA, however, shows that now TA no longer holds the better acceptance criterion. The main reason for this might be the fact that TA accepts any change for the worse as long as it is sufficiently small, whereas in SA such changes might also be rejected. This implies that under TA, solutions might slowly but persistently drift into the wrong direction. The introduction of the elitist principle improves the situation by assigning the challenged agent the best solution found so far. The "standard" version of the MA, using SA's Metropolis function, on the other hand reports very stable results with and without elitists. Though the elitist principle sometimes seemingly freezes the agents in a local optimum, it also prevents them from being too far from the global optimum: agents representing local optima that are worse than the elitist's solutions will eventually be replaced with the elitist and can then no longer pass their inferior local solution on to their neighbors. Hence, the versions with elitists report solutions with higher average objective values than those without for the ETL^{emp} constraint.

The computationally least demanding of the compared risk constraints are those that assume normally distributed returns. Here, the same effects as for the models with the ETL^{emp} constraint can be seen, yet with an even smaller bandwidth for the reported results: the average ratio between the optimum and the median of the reported solutions is just 1.0001. The TA criterion has a slightly higher chance of missing the optimum, which can be improved with the elitist principle. More stable results come from the versions with the Metropolis function: virtually never a

---

[24] ETL satisfies a number of desirable properties of risk measures and therefore is a "coherent risk measure" as defined by Artzner, Delbaen, Eber, and Heath (1999) whereas VaR isn't; see, e.g., Pflug (2000).

result below the (supposed) global optimum is reported, especially when the elitist principle is included: the elitist strategy improves the results with a high statistical significance. However, given the actual magnitude of these improvements as well as the differences between results from SA and MA, these advantages are rather of academic interest, but can be more or less neglected in practical application.

Summarizing, the version with TA tends to find solutions that are at or relatively close to the (supposed) global optimum. This is quite advantageous, when the problem space is very rough (and the results found with SA tend to be less stable), yet less favorable when the solution space is rather smooth and the standard version of MA has hardly any problem of finding the (supposed) global optimum.

As indicated earlier, the elitist strategy tends to amend both local and global search aspects: at early iterations this principle might well keep agents from moving too far into the wrong direction, in particular when their neighbors on the ring "got lost" themselves, too. At later iterations it retrieves agents from what is likely to be a local optimum that could hardly be escaped otherwise. In preliminary tests, the elitist strategy turned out to be quite helpful as it appears to make the heuristic less sensitive towards parameters such as cooling factor and initial temperature (or finding the threshold sequence) as well as the range of modifications within the local search. Once all the parameters have undergone the necessary fine tuning and the (local) search procedures are made more sophisticated, their remaining advantage (and disadvantage) is merely the higher convergence speed: With the elitist strategy, the heuristic tends to find the eventually reported solution at an earlier stage than do the versions without elitists as the "news" of found good solutions is spread faster – with a small risk of quickly trapping the population at a local optimum.

6.4   Results for Stock Portfolios

6.4.1   Assumed and Resulting Distribution

Working with empirical distributions usually increases the number of local optima and makes the optimization problem significantly harder than it would be with normally distributed returns. It therefore appears reasonable to check whether the additional efforts are worth the trouble, i.e., whether the actual returns are normally distributed or not. As effects due to the optimization process are to be excluded at this stage, 30 000 portfolios are randomly generated as described above which are not optimized but assigned random, non-negative asset weights that add up to 1. Performing a standard Jarque-Bera test, the normality assumption can be rejected both in sample and out of sample for approximately 75% of the assets and for more than 67% of the random portfolios.[25] The reasons for this large number of rejections are mainly days with extreme gains or losses, leading to skewness and excess kurtosis. Hence, in a Kolmogorov-Smirnov test, which is more "tolerant" to outliers, the rejection rate reduces to about 28% for the assets. Both tests indicate that the normality assumption for the assets' returns might lead to specification errors. For portfolios with random weights, however, the Kolmogorov-Smirnov test rejection rate for the normality assumption becomes as low as 2%.

For each of these random portfolios (as well as for the included 25 stocks per portfolio), the VaR was estimated both under empirical (VaR^{emp}) and under normal distribution (VaR^{norm}) for a shortfall probability of $\alpha = 0.10$. Table 6.2 summarizes how often the estimated VaR^{emp} and VaR^{norm}, respectively, were actually exceeded in the consecutive out of sample period. The results indicate that the empirical distribution appears well apt to estimate the assets' as well as the portfolios' VaR: in particular for shorter out of sample periods, losses beyond the estimated VaR occur in just slightly more than 10% of all cases which corresponds well with the (accepted and expected) shortfall probability of $\alpha = 0.10$. Under the normal distribution, however, the risk appears to be overestimated and the actual shortfalls beyond the VaR limit occur notably less often than the chosen level of α would predict.

When the asset weights are no longer random but actually optimized according to the introduced selection problem, the resulting portfolios appear to have similar properties – yet only at first sight. Table 6.3 summarizes the test statistics for the assets' and the optimized portfolios' returns. Based on a Jarque-Bera test, the hypothesis of normal distribution can be rejected for approximately three quarters of all data series of available individual assets. This ratio is virtually the same for in sample asset returns as well as out of sample returns (for the next 200 trading days) and it is not really affected if assets with zero weights are excluded; it therefore corresponds well to the results for the random weight portfolios. As this test is based on the observations' skewness and kurtosis, this might indicate that none of the risk

---

[25] For all statistical tests, the level of significance is 5%.

T = length of out of sample period	assets		portfolios with random weights	
	VaR^{emp}	VaR^{norm}	VaR^{emp}	VaR^{norm}
1 day	.1037	.0850	.1002	.0870
10 days	.1033	.0847	.1001	.0871
20 days	.1036	.0850	.1003	.0873
50 days	.1045	.0860	.1014	.0881
100 days	.1056	.0871	.1026	.0893
200 days	.1080	.0895	.1050	.0916

Tab. 6.2: Frequency of out of sample losses exceeded the estimated VaR ($r \leq r_{VaR}$) for $\alpha = 0.10$

VaR^{emp}	in sample				out of sample (200 trading days)			
	ETL^{emp}	VaR^{norm}	ETL^{norm}	VaR^{emp}	ETL^{emp}	VaR^{norm}	ETL^{norm}	
Jarque-Bera test								
available assets	.741	.741	.741	.741	.755	.755	.755	.755
included assets	.744	.747	.738	.738	.754	.754	.757	.757
optimized portfolios	.780	.660	.524	.524	.648	.652	.628	.628
Kolmogorov-Smirnov test								
available assets	.296	.296	.296	.296	.283	.283	.283	.283
included assets	.313	.354	.345	.345	.260	.264	.262	.262
optimized portfolios	.560	.040	.016	.016	.028	.028	.028	.028

Tab. 6.3: Fraction of assets and optimized portfolios for which the hypothesis of normal distribution can be rejected at the 5% level

constraints leads to a different rate of inclusion or exclusion of assets with certain characteristics of their higher moments.

A closer look at the resulting portfolios, however, reveals first effects of the different distribution assumptions. Whereas the Jarque-Bera test rejects the normality assumption for a fraction of 0.7 of the random weight portfolios, assuming the normal distribution in the risk constraint, VaR^{norm}, appears to enforce normally distributed portfolio returns whereas VaR^{emp} has the opposite effect: here, the rejection rates are 52.4% and 78%, respectively. With the Kolmogorov-Smirnov test, these particularities of portfolios optimized under VaR^{emp} become even more evident. It is just some 29% of the available assets' data series for which normality can be rejected (both in and out of sample) and slightly more of the actually included assets that have non-normally distributed in sample returns; these fractions are more or less independent of the risk constraint and are also in line with the random weight portfolios. The resulting portfolios, however, differ massively: Under VaR^{emp}, more than half of the optimal portfolios are not normally distributed according to the Kolmogorov-Smirnov test, whereas for all other portfolios the rejection rate is drastically lower: for those under ETL^{emp}, it is 4%, and for the random weight portfolios as well as those optimized under normality it is just about 2%. This corresponds with the results for the random weight portfolios.

These results imply that the VaR^{emp} constraint produces portfolios with an in sample distribution that is far from normal. However, this cannot be attributed to a hidden preference for assets with fancy distributions: as argued previously, the included assets do not differ noticeably from the available ones. When looking at the out of sample distributions, these particularities seem to largely vanish again, leading to a decreasing number of portfolios for which the normality assumption can be rejected. For the VaR^{norm} constrained portfolios, the situation appears to be the other way round: in sample, more portfolios seem to have normally distributed returns than out of sample, as the rejection is higher for the out of sample data. This effect, however, is considerably smaller.

6.4.2   Great Expectations and Broken Promises: The Resulting Stock Portfolios

It is a well-known problem that under a VaR regime the optimization process does not really bother about extreme losses and high kurtosis as long as they occur with a probability of less than α (i.e., do not interfere with the VaR constraint). The magnitude of the excess loss does not count in the risk measure, and if extreme losses occur relatively seldom, they do not have a strong effect on the expected utility either; they might even be preferred when they are paralleled with extreme gains. In a volatility framework, on the other hand, these extreme losses do show in the risk measure and are therefore accepted less easily.

This effect can be found in our results for the optimized VaR^{emp} portfolios where for each portfolio, a higher return is expected than if the same portfolio would have been optimized under VaR^{norm} (see Figure 6.3(a)). A closer look at the portfolios' statistics also shows that VaR^{emp} constrained portfolios have both higher expected

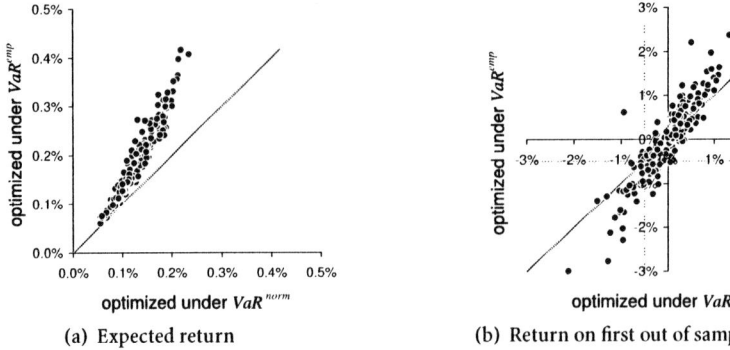

(a) Expected return   (b) Return on first out of sample day

Fig. 6.3: Counter plots for portfolios optimized under different distribution assumptions (gray solid: 45° line; gray dashed: risk constraint $r_{VaR} = -0.5\%$)

returns as well as higher volatility. Taking into account the large number of these portfolios that have not normally distributed returns (i.e., mean and volatility are not sufficient to characterize their distribution), measuring the VaR of these portfolios is likely to depend heavily on the assumed distribution. This problem becomes apparent when a once optimized portfolio is evaluated with the same risk constraint yet with a different distribution assumption: whereas for the empirical distribution, the condition $r_{VaR}^{emp} = -0.5\%$ is perfectly met by constraint, a VaR estimation for these portfolios according to $r_{VaR}^{norm} = E(r) - |u_\alpha| \cdot \sigma$ leads to values ranging from -0.57% to -1.24%. These differences between r_{VaR}^{emp} and r_{VaR}^{norm} show that evaluating a VaR^{emp} constrained portfolio with a mean-variance approach leads to a considerable gap between results – a gap that has merely been widened by the use of the empirical distribution in the optimization process as it is significantly larger than for portfolios with random weights.

Optimizing under the normality assumption does not lead to such a discrepancy: If a VaR^{norm} constrained portfolio is evaluated with the empirical distribution by determining the $(\alpha \cdot T)$-th worst of the T observations, the results range from -0.32% to -0.56%, i.e., are close to and on either side of the accepted value of $r_{VaR} = -0.5\%$ (which, by constraint and assumed distribution, is equal to r_{VaR}^{norm}). Despite its known specification errors, the normality assumption does not open a similarly large gap between the VaR estimates with different distribution assumptions as has been by using the empirical distribution in the optimization process.

The fact that VaR^{emp} constrained portfolios have non-normal distributions does not imply that the use of the empirical distribution is necessarily superior to the models with normal distribution. The downside of VaR^{emp}'s being able to respond to peculiarities of the in sample data (and actually doing so) becomes evident when the out of sample performance is analyzed. Figure 6.3(b) illustrates that the returns on the first out of sample day differ significantly depending on the distribution assumption: When optimized under the empirical distribution, the (extreme) gains tend to be higher as they are for the same portfolio when optimized under the normality assumption, yet so are the (extreme) losses.

This mismatch between in sample and out of sample results is not just observable on the first out of sample day, but can also be observed for longer out of sample periods (see Table 6.4). The original risk constraint demands that in a fraction α of all trading days the returns must not fall below $r_{VaR} = -0.5\%$; portfolios optimized under the VaR^{emp} constraint, however, actually generate returns of about -1% or even less in $\alpha \cdot T$ of the $T = 10, 20, 50, 100$, and 200 trading days following the in sample period. The incurred VaR is therefore at least twice the original limit. The actual VaR is therefore drastically underestimated by the empirical distribution – yet would have been announced by these portfolios' volatilities which would have predicted a significantly higher VaR than the empirical distribution. Given these results, it is not surprising that VaR^{emp} constrained portfolios also violate the original limit of r_{VaR} too often: the given risk limit of $r_{VaR} = -0.5\%$ is exceeded in 20.7% up to 23.6% of the out of sample days, which is more than twice the specified value of $\alpha = 0.1$.

Out of sample gains and losses of portfolios optimized under the normality assumption are significantly closer to the original limit. Here, the actual return realized in α of the out of sample days is about -0.6%. The estimated r_{VaR} is exceeded more often than assumed by α and as had been for the random weight portfolios; yet the deviations between the expected and the actually incurred frequency is significantly smaller under the normality assumption than under empirical distributions. These differences can partially be attributed to the specification errors of the normal distribution. The results from additional computational studies for this data set show that more sophisticated parameter estimations (such as estimating the

| T = length of out of | return on $(\alpha \cdot T)$-th worst day | | frequency of $r \leq r_{VaR}$ | |
of sample period	VaR^{emp}	VaR^{norm}	VaR^{emp}	VaR^{norm}
1 day	—	—	.2360	.1480
10 days	−1.28%	−0.83%	.2132	.1336
20 days	−1.10%	−0.69%	.2164	.1308
50 days	−0.99%	−0.64%	.2070	.1270
100 days	−1.00%	−0.64%	.2138	.1357
200 days	−1.00%	−0.64%	.2176	.1388

Tab. 6.4: Out of sample results for portfolios optimized under different distribution assumptions and VaR with $r_{VaR} = -0.5\%$ and $\alpha = 0.10$

volatility with a GARCH model) can improve the reliability and lead to smaller differences between the predicted and the actual frequency of shortfalls. The quality of the predictions can also benefit from the use of alternative parametric distributions, as suggested, e.g., by Buckley, Comezaña, Djerroud, and Seco (2003) or de Beus, Bressers, and de Graaf (2003). Also, own experiments with alternative parametric distributions such as the S_U-normal[26] indicate that there are better choices than the standard normal distribution. However, as this study is merely concerned with the effects of empirical distributions, a more detailed discussion of the results on parametric methods would go beyond its scope.

As argued before, VaR^{emp} portfolios have not just higher expected returns than their VaR^{norm} counterparts, they also have higher volatilities. VaR does not measure the magnitude of losses exceeding the r_{VaR} limit, but only their frequency. This provokes tailor-made solution under the empirical distribution and allows manipulating the weights such that the in sample days with returns close to the critical value, r_{VaR}, are mostly days where the returns are actually slightly above r_{VaR} and therefore do not count towards the VaR criterion. Hence, when the VaR limit is exceeded, the failure tends to be high, but the optimization process ignores that the higher expected return comes at the expense of higher volatility. Out of sample, this "trick" of avoiding slight violations of the r_{VaR} constraints will not hold any longer and days with returns close to r_{VaR} will include days above and below r_{VaR}. As shown previously, the actual out of sample distribution will be much closer to the normal

---

[26] See Johnson (1949).

T = length of out of	risk constraint				random
of sample period	VaR^{emp}	ETL^{emp}	VaR^{norm}	ETL^{norm}	weights
10 days	.044	.060	.080	.080	.047
20 days	.068	.072	.072	.072	.049
50 days	.100	.088	.100	.100	.050
100 days	.272	.168	.176	.176	.067
200 days	.496	.240	.248	.248	.085

Tab. 6.5: Fraction of portfolios with significantly different in sample and out of sample distributions at a 5% level of significance based on a Kolmogorov-Smirnov test

distribution as the in sample distribution would suggest. Hence, out of sample the high volatility will show up in the increased number of days where the original VaR limit is violated as well as the higher losses in the $(\alpha \cdot T)$ worst days. The longer the out of sample period, the more obvious these differences between the in sample and out of sample distributions become (see Table 6.5). Figures 6.4(a) and 6.4(b) illustrate this undesired feature for one typical specimen portfolio.

This, however, is also true for the VaR^{norm} constrained portfolios, though to a far lesser degree. The differences between in sample and out of sample distributions are more distinct when compared to portfolios with random weights, hence the optimization process under the normality assumption, too, has an effect on the resulting distributions' properties. Compared to the VaR^{emp} constrained portfolios, however, the differences are by magnitude smaller (see Table 6.5). More important, the data fitting around the r_{VaR} limit is largely avoided when the optimization process is based on the parametric distribution (see Figures 6.4(c) and 6.4(d)). As indicated above, more sophisticated parametric approaches might have additional beneficial effects.

When ETL is the risk constraint of choice, the differences in the optimized portfolios due to the assumed distribution are observable, yet not very distinct. Here, optimizing under the assumption of normally distributed returns leads to out of sample statistics that are closer to the values originally aimed at, while the use of the empirical distribution produces slightly better in sample and slightly worse out of sample results. As the ETL criterion does look at the magnitude of the losses when r_{VaR} is exceeded, accepting high losses becomes less attractive, and so does avoiding

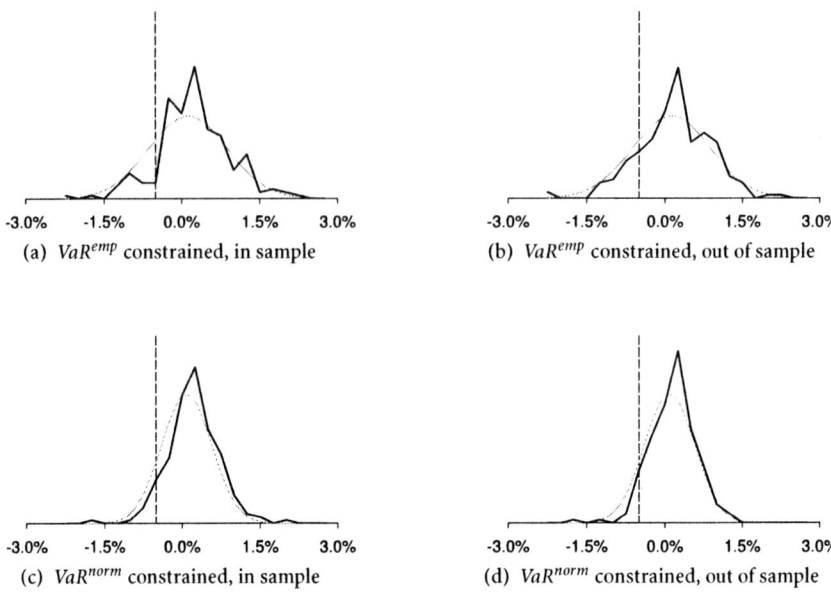

Fig. 6.4: *Effect of the risk constraint on the return distribution for specimen portfolio No. 87 (gray lines: standard normal distribution based on the respective in sample mean and variance; straight vertical line at $r_{VaR} = -0.5\%$)*

slight violations of the critical value, r_{VaR}. The empirical distribution's "advantage" of tailoring the solution is strongly reduced, and the distribution assumption has no noticeable consequences.

Under the assumption of normally distributed returns, the decision of whether to use ETL or VaR does not matter any more. As can be seen from the myopic optimization model, the constraints differ only in the weight on the portfolio's volatility, namely $|u_\alpha|$ and $\phi(u_\alpha)/\alpha$ for VaR^{norm} and for ETL^{norm}, respectively. As both weights are constant, the risk constraint effects the amount, $(1-q)$, invested into the safe asset rather than the risky portfolio itself. Nonetheless, the fraction of portfolios with different in sample and out of sample distribution is noticeably higher than for random weight portfolios, indicating that here, too, some undesired data-fitting might have taken place. With this study's focus on empirical distributions and VaR, however, this issue has to be left to future research.

In addition to the presented results, additional computational studies have been performed with alternative numbers of assets included in the portfolios and with alternative critical values for α and r_{VaR}. The main findings of the presented results, however, remain unaffected: Under VaR^{emp} the optimized portfolios promise higher returns and utility, but out of sample violate the risk constraint distinctly more often than their VaR^{norm} counterparts, and these differences are reduced under ETL.

6.5 Results for Bond Portfolios

6.5.1 Assumed and Resulting Distribution

The decision of whether to estimate the VaR with the normal (or any other parametric) rather than the empirical distribution depends on how well the main properties of the observed data for the assets (or at least, via the CLT, the resulting portfolios) can be captured with the parametric distribution. For the given data set, the portfolio values appear far from normally distributed: regardless of the method for VaR estimation, there is hardly any optimized portfolio where the null hypothesis of normal distributed price changes cannot be rejected at the usual 5% level of significance both based on a standard Jarque-Bera test (as can be seen in Table 6.6) and the Kolmogorov-Smirnow test. Looking at the bond prices the null is rejected for virtually any of the assets in the data set – the details can therefore be omitted in the sense of brevity. The main reasons for the high rate of rejection are the leptokurtic and highly peaked distributions in the portfolios: even when taking into account that the higher moments do not necessarily exist (and therefore the Jarque-Bera test, using skewness and kurtosis, might not be appropriate) and calculating the Selector statistics[27] the picture remains more or less unchanged (see Table 6.7).

At first sight, this seems to confirm the view that the normality assumption in the optimization process might be inadequate and that the use of empirical distributions might be the better choice: For most of the portfolios (see Table 6.8) and for an even higher share of the included assets, the hypothesis of same in and out of sample distributions cannot be rejected, hence using past realizations for estimates of future outcomes appears legitimate.

---

[27] See Schmid and Trede (2003).

	method	N = 10, α = ...			N = 20, α = ...		
		2.5%	5.0%	10.0%	2.5%	5.0%	10.0%
in sample	empirical	.984	.988	.988	.984	.984	.996
	empirical, weighted	.984	.992	.992	.988	.992	.992
	normal	.984	.988	.988	.984	.984	.988
	normal, weighted	.984	.988	.984	.988	.988	.988
out of sample	empirical	.904	.896	.912	.900	.908	.916
	empirical, weighted	.896	.912	.896	.908	.900	.904
	normal	.900	.900	.908	.884	.896	.912
	normal, weighted	.912	.908	.900	.908	.908	.908

Tab. 6.6: Fraction of portfolios for which the normality assumption can be rejected (Jarque-Bera test, 5% significance)

	method	N = 10, α = ...			N = 20, α = ...		
		2.5%	5.0%	10.0%	2.5%	5.0%	10.0%
in sample	empirical	.976	.988	.988	.968	1.000	1.000
	empirical, weighted	.988	.992	.988	.984	1.000	1.000
	normal	.988	.988	.988	1.000	1.000	1.000
	normal, weighted	.988	.988	.988	1.000	.996	1.000
out of sample	empirical	.948	.960	.960	.924	.944	.956
	empirical, weighted	.940	.956	.964	.940	.936	.936
	normal	.944	.952	.944	.924	.928	.916
	normal, weighted	.948	.956	.952	.944	.924	.940

Tab. 6.7: Fraction of portfolios for which normal distribution can be rejected (Selector Statistics test for leptokurtosis, 5% significance)

method	N = 10, α = ...			N = 20, α = ...		
	2.5%	5.0%	10.0%	2.5%	5.0%	10.0%
empirical	.200	.236	.204	.288	.324	.304
empirical, weighted	.184	.204	.200	.296	.308	.320
normal	.204	.220	.220	.300	.296	.300
normal, weighted	.204	.204	.220	.288	.292	.312

Tab. 6.8: Fraction of portfolios were the H_0: same in sample and out of sample distributions (100 out of sample days) can be rejected at the 5% significance level (Kolmogorov-Smirnov test)

method	$N = 10, \alpha = ...$			$N = 20, \alpha = ...$		
	2.5%	5.0%	10.0%	2.5%	5.0%	10.0%
empirical	2.6 %	4.2%	9.1%	3.2%	5.6%	10.2%
empirical, weighted	2.4%	4.1%	8.2%	3.0%	5.4%	9.9%
normal	2.9%	4.1%	6.4%	3.2%	5.3%	8.1%
normal, weighted	2.8%	4.0%	6.2%	3.1%	5.1%	8.0%

Tab. 6.9: Percentage of portfolios with random asset weights exceeding the estimated VaR limit on the first out of sample day for the two case sets with a confidence level of α

To test whether the distributions are stable and allow reliable estimates of the VaR, we repeatedly generated random weights for any bundle of assets in the two case sets where the integer and the budget constraints are the only restrictions. Then, the share of portfolios with out of sample losses higher than the expected VaR is determined. As can be seen from Table 6.9 for the first out of sample day, the use of the empirical distributions allows for estimations of the VaR such that the frequency of larger losses corresponds more or less to the respective confidence level. Under the normality assumption, higher values for α result in overly cautious estimations of the VaR – violations of which occur less often than expected. In particular for higher values of α, the empirical distribution produces more reliable results than the normal distribution. This relative advantage remains unaffected when longer out of sample periods are used for evaluation. For smaller values of α, the advantage of the empirical distribution is less apparent; with respect to the number of in sample observations, the shortfall probability of $\alpha = 2.5\%$ refers to the worst 5 observations, estimates of r_{VaR} are therefore more difficult – and so are comparisons of the differences between empirical and parametric distributions. In the light of the above discussion about the need for more observations when α is low, the results for low α's will be reported for the sake of completeness, yet the main discussion will focus on $\alpha = 10\%$ where there are more observations for returns at or below the VaR limit and the presented conclusions are all supported by the usual levels of statistical significance.

method	$N = 10, \alpha = ...$			$N = 20, \alpha = ...$		
	2.5%	5.0%	10.0%	2.5%	5.0%	10.0%
empirical	5.2%	7.6%	16.8%	8.8%	10.8%	16.9%
empirical weighted	6.0%	8.0%	15.2%	8.4%	10.8%	16.1%
normal	3.2%	3.6%	7.2%	4.4%	6.8%	8.4%
normal weighted	3.2%	4.0%	6.0%	3.6%	5.6%	6.8%

Tab. 6.10: Percentage of optimized portfolios exceeding the estimated VaR, $V_0 \cdot E\left(\delta^{VaR}\right)$, on the first out of sample day for the two case sets with a confidence level of α

6.5.2 The Hidden Risks in Optimized Bond Portfolios

Unlike portfolios with random weights, the value of portfolios that are optimized under the empirical distribution will fall significantly more often below $V_0 \cdot E\left(\delta^{VaR}\right)$, the expected VaR[28], than the chosen confidence level α. On the first out of sample day (Table 6.10) the actual frequency of excessive shortfalls will be 1.5 to three times the frequency originally expected (depending on α and case set). When the same portfolios are optimized under the normal distribution, however, the frequency will be underestimated only for small α's, for high confidence levels, on the other hand, the frequency will be overestimated, i.e., the VaR is estimated too cautiously. The assumption of the normal distribution leads (for both optimized and random portfolios) to more cautious estimates of the VaR when α is high. The extreme leptokurtosis of the actual distributions cannot be captured by the normal distribution, and as a result it is hardly possible to get reliable estimates for the VaR limit: For large confidence levels of α, the VaR limit is estimated too far away from the expected value, for low confidence levels.[29]

The smaller α, the more only extreme outliers contribute to the shortfalls – the estimated frequencies for the first out of sample day are therefore more sensible to the chosen sample. Table 6.11 therefore takes into account larger out of sample

---

[28] Due to the specification and the chosen assets, the critical VaR, the out of sample data were compared to, is set to $E\left(\delta^{VaR}\right) \leq \underline{\delta}^{VaR}$, the loss actually expected with the planed probability of α.

[29] For small values of α the opposite can be observed: the VaR is underestimated, and the limit is violated too often. With respect to the data set, however, tests with smaller values of α than the ones presented were not possible, a more detailed discussion of these effects has therefore be left to future research.

	method	$N = 10, \alpha = \ldots$			$N = 20, \alpha = \ldots$		
		2.5%	5.0%	10.0%	2.5%	5.0%	10.0%
$T_{oos} = 50$	empirical	7.3%	9.9%	17.8%	9.8%	13.5%	19.7%
	empirical weighted	7.2%	9.8%	16.4%	9.3%	12.9%	18.8%
	normal	5.6%	6.8%	8.9%	7.2%	8.9%	11.6%
	normal weighted	5.3%	6.4%	8.5%	6.8%	8.4%	11.0%
$T_{oos} = 100$	empirical	7.8%	10.6%	18.6%	10.4%	14.0%	20.1%
	empirical weighted	7.8%	10.5%	17.2%	10.1%	13.4%	19.2%
	normal	6.2%	7.5%	9.6%	8.0%	9.6%	12.1%
	normal weighted	6.0%	7.2%	9.3%	7.7%	9.2%	11.6%

Tab. 6.11: Average percentage of the first T_{oos} out of sample days where the loss exceeds the expected VaR for optimized portfolios

periods, namely the first 50 and 100 out of sample trading days for the $N = 10$ and the $N = 20$ case sets, respectively. The basic conclusion from the first out of sample day that has been drawn for the "empirically" optimized portfolios, however, remains unchanged: the actual percentage of cases where the VaR is violated is significantly higher than the accepted level of α. For the optimization results under the normal distribution, the frequencies of excessive shortfalls increase; resulting figures closer to α when α is large, yet exceeding it significantly when α is low. In the light of the results from the previous' sections, one can conclude that the specification errors of the normal distribution become more obvious in these cases.

The advantage of the empirical over the normal distribution that had been identified for non-optimized portfolios and the statistical properties of the actual distribution, seems therefore lost and in some cases even reverted into the opposite when a VaR constraint is used in the optimization process. Despite its specification errors, the normal distribution seems to cause fewer problems than using empirical distributions that has been shown to be closer to reality for the single assets and non-optimized portfolios.

The major reason for this is that VaR is a quantile risk measure and therefore focuses on the number of shortfalls rather than their magnitude.[30] This can and will be exploited when empirical distributions are used. When optimizing under

---

[30] See also Artzner, Delbaen, Eber, and Heath (1999).

an empirical distribution, a number of excessive losses beyond the specified VaR limit will contribute equally to the confidence level α as would the same number of small losses; the optimization process will therefore "favor" losses that come with high yields. Since it is usually the high yield bonds that exhibit massive losses in the past, these bonds will be given high weights. The problems arising from this effect are reinforced when the high yield of a bond comes from a small number of high losses rather than several small losses: a loss beyond the specified VaR limit will be considered a rare event, and the loss limit estimated with the confidence level α will be distinctly below the accepted limit, i.e., $E\left(\delta^{VaR}\right) \ll \underline{\delta}^{VaR}$. Out of sample, this expected limit might turn out to be too optimistic and is therefore violated too often, hence the observed percentage of days with out of sample losses beyond the expected VaR is distinctly higher than the originally accepted level of α.

In addition, there is a hidden danger of data fitting for the empirical distribution: Slight in sample violations of the specified VaR limit of $\underline{\delta}^{VaR}$ can (and will) sometimes be avoided by slight changes in the combination of assets' weights that have only a minor effect on the portfolio yield. As a consequence, there might be more cases close to the specified VaR than the investor is aware of since they are just slightly above the limit and therefore do not count towards the level α; out of sample, however, this hidden risk causes more shortfalls than expected.[31]

Both effects become more apparent from the scatter plots in Figure 6.5 where the results for portfolios optimized under empirical distributions are directly compared to the results when optimized under normal distribution. The magnitude of extreme losses shows up when the risk is measured in terms of volatility: "empirical" portfolios accept a standard deviation of up to CHF 20 000 and, on rare occasions, even more. When optimizing under the normality assumption, the definition of VaR imposes an implicit upper limit on the volatility of $\sigma_{V_\tau} \leq \frac{E(V_\tau)-V_0 \cdot \left(1-\underline{\delta}^{VaR}\right)}{u_\alpha}$, which, for $\alpha = 0.1$, is below CHF 10 000 for any portfolio in the case set. The volatility will be (approximately) the same regardless of the assumed distribution only if the "normal" portfolios have low volatility; when the optimal portfolios under normality actually make use of the specified risk limit (in sample), then their empirical coun-

---

[31] Because of the peakedness of the distributions of bond returns and the discussed effect, that $E\left(\delta^{VaR}\right) < \underline{\delta}^{VaR}$ for larger values of α, this effect of data fitting does not show as often as for assets with other empirical distributions such as stocks; see Maringer (2003a).

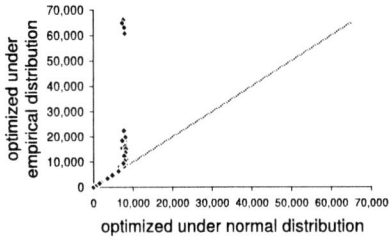
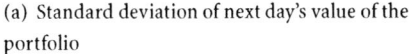
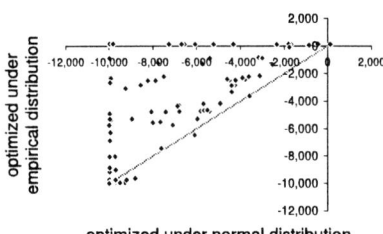

(a) Standard deviation of next day's value of the portfolio

(b) Loss expected with a probability of α

Fig. 6.5: Expected standard deviation and VaR of optimized portfolios for the $N = 20$ case set with $\alpha = 0.1$ (gray: $45°$ line; gray dashed: risk constraint at CHF $-10\,000$)

terparts are very likely to accept large variations in the respective portfolio's value (see Figure 6.5(a)).

Figure 6.5(b) illustrates that, at the same time, there is a considerable number of cases where, when optimized under empirical distributions, the portfolios have smaller expected loss than one would expect from a corresponding portfolio with the same assets yet optimized under normal distribution and therefore different asset weights. When α is chosen rather large, the peakedness of the empirical distribution results in a VaR limit closer to the portfolio's expected value than predicted when the normal distribution is assumed: the rare, yet extreme in sample losses are perfectly ignored by the empirical distribution. If these extreme losses are rare enough, it might even happen that given a sufficiently large confidence level the estimated VaR limit will be a gain rather than a loss. This can be observed already for some portfolios in the $\alpha = 0.1$ case. Under the normal distribution, on the other hand, they do show up. Under empirical distributions, the investor will therefore be more inclined to accept extreme (in sample) losses without violating the risk constraint in sample; under the normal distribution, the investor will be more reluctant. This explains why a portfolio optimized under the empirical distribution will have a higher expected yield than a corresponding portfolio containing the same assets yet optimized under the normality assumption. Figure 6.6(a) illustrates these differences for portfolios with $N = 20$ bonds. The larger the set of available assets, the more is the investor able to make use of this fact. Not surprisingly, the deviations between accepted α and actual percentage of out of sample shortfalls therefore in-

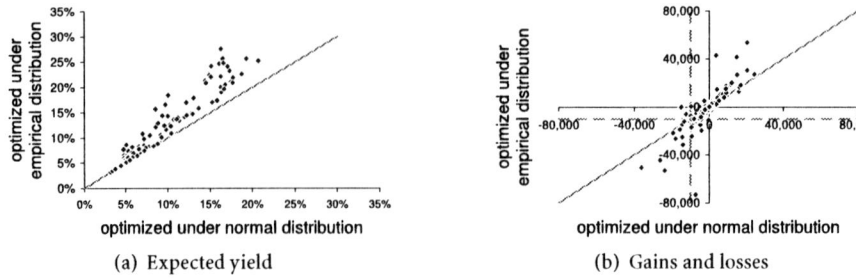

Fig. 6.6: Expected portfolio yield (per annum) and gains & losses on the first out of sample day depending on the distribution assumption for the N = 20 case set and α = 0.1 (gray: 45° line)

crease, when N is larger, i.e., the investor has a larger set of alternatives to choose from (see Tables 6.10 and 6.11).

The consequences of these effects are twofold: First, the "empirical" optimizer underestimates the chances for exceeding the VaR limit since the scenarios where the limit is narrowly not exceeded in sample have a fair chance of exceeding it out of sample – hence the percentage of cases or days with losses beyond $E\left(\delta^{VaR}\right) \cdot V_0$ is higher than α, i.e., the expected percentage. Second, since the "empirical" optimizer does accept extreme losses in sample, she has a good chance of facing them out of sample as well. The "empirical" investor will therefore not only encounter losses exceeding the estimated VaR limit more frequently than the "normal" investor, the "empirical" investor's losses will also be higher, as can be seen from Figure 6.6(b).

To what extent the deficiencies of empirical distributions are exploited in the op-timization process depends on several aspects where the number of the in sample observations or simulations certainly is a very crucial one. Long time series, how-ever, are not always available nor can they be reliably generated.[32] In addition the stability of the distribution becomes a major issue. Including more historic data might bring only diminishing contributions when weighted values (or alternative prediction models such as GARCH models) are used. Detailed tests of these aspects, however, were not possible with the available data and must therefore be left to fu-ture research.

---

[32] The problem of small sample sizes becomes even more apparent in practical when credit portfolios or non-publicly traded assets are considered instead of publicly traded assets.

6.6   Conclusion

The main findings from the empirical studies in this chapter are threefold: (i) for the used data the assumption of normal distribution can mostly be rejected and the empirical distribution is superior to the normal distribution; (ii) Value at Risk is a reliable measure of risk, in particular when used in conjunction with empirical distributions – and (iii) the opposite of the previous two points becomes true when Value at Risk under empirical distributions is used as an explicit constraint on portfolio risk in an optimization setting. Though empirical distributions are superior in describing the assets return and potential losses, this advantage is destroyed when included in the optimization process.

The reason for this can be found in the underlying concept: Value at Risk focuses on a single point of the distribution of returns or prices, namely the loss that will not be exceeded with a certain probability, but does not directly account for the magnitude of losses exceeding this limit. When VaR is estimated via empirical distributions based on historic simulation, as is done in this study, or Monte Carlo simulations, extreme losses might be ignored and assets, exhibiting these extreme losses, will readily be included in the portfolio when they have a sufficiently high expected return. Also, the use of empirical distributions encourages the optimizer to manipulate the asset weights in a way that losses close to the VaR limit are just small enough so that this limit is not exceeded and these losses will not count towards the shortfall probability. Both effects lead to severe under-estimating the actual risk of the portfolio.

Even when asset returns are not normally distributed, the assumption of this parametric distribution leads to more reliable risk estimations because extreme losses enter the optimization process via the volatility. When VaR is considered in the optimization process, parametric distributions might therefore be superior, despite the fact that their imprecision in measuring risk lead to imprecise estimates.

The results from this study suggest that empirical distributions should be used very reluctantly in VaR optimization, yet also that more research on the use of parametric distributions appears desirable.

Chapter 7

Finding Relevant Risk Factors in Asset Pricing

7.1   Introduction

Explaining past stock returns and reliably predicting future performance has been a major issue in the finance literature ever since. Meanwhile, several theoretically sound and well-founded equilibrium models exist, arguably the most popular of these being William Sharpe's single index model, the *Capital Asset Pricing Model* (*CAPM*).[1] The CAPM estimates an asset's risk premium according to its contribution and relation to the market risk. The main reason for the popularity of the CAPM, its simplicity, has also given reason for critique, as relying on a single factor, the market index, might leave too much of an asset's risk unexplained. Multi-index models such as Stephen A. Ross's *Arbitrage Pricing Theory* (*APT*),[2] on the other hand, usually follow the intuitively appealing idea that asset prices are mainly driven by several factor prices that, ideally, have some fundamental and plausible relationship to the underlying company. In this case, deviations between an asset's actual return and its expected return can be largely traced back to shifts in the corresponding factors.

The basic version of the APT, however, does not necessarily demand causal dependencies between factors and asset prices: it derives its indices solely on statistical

---

[1] For a detailed presentation of the original 1964 paper and follow-up literature, see, e.g., Sharpe, Alexander, and Bailey (2003). A short introduction can be found in section 1.2.2.

[2] See, e.g., Ross (1976) and Ross (1977) as well as Roll and Ross (1980). A presentation of relevant literature can be found in Shukla (1997). See also section 1.2.4.

grounds from the available asset prices in a complete market. Hence, neither the resulting factors nor their weights for the priced asset have an obvious (economic) meaning – nor need they have to have one as this approach is not meant to identify economic fundamentals. Given a complete market where the number of possible states equals the number of available, linearly independent assets from which factors can be derived, the APT replicates any asset by finding "portfolios" of indices (or factors) that generate the same state dependent payoffs as the asset. The asset's fair price can therefore be determined via the *law of one price* as the prices of the indices are known. When the market is not complete, then the security's return can still be expressed with a linear regression model (LRM) though in this case there remains some residual risk.[3]

In practice a perfect *ex post* (let alone, *ex ante*) replication of payoffs or returns is hardly possible: deriving the market specific indices needed for the theoretical version of the APT is not possible, and readily available factors have to be used instead. Applying the APT therefore involves finding an acceptable trade-off between model size and model reliability by selecting a (preferably small) set of factors that captures as much of the asset's price changes as possible. The problem of factor selection is therefore crucial for the successful application of the APT. The literature offers two basic types of factor selection: The first group chooses factors that are assumed to have some economically plausible and fundamental relationship to the considered asset, whereas the second group merely relies on statistical analyses.

The use of fundamental factors demands either detailed knowledge of the firm and the availability of suitable factors on an industry level (which usually differ between firms), or the use of general economic factors that are not separately selected for individual assets but work across firms, industries and markets. The use of such economically plausible factors aims to find long term relationships and explanations for price movements. Most of the work so far has focused on macroeconomic factors, most prominent the results in Chen, Roll, and Ross (1986) and Burmeister and Wall (1986). This is largely in line with, e.g., the recommendations by BIRR® Portfolio Analysis, Inc., a consulting company founded by finance professors and APT-theorists, Edwin Burmeister, Roger Ibbotson, Stephen Ross, and Richard Roll, who apply the APT in their risk models. The focus of their models is on "unexpected

---

[3] Quantitative and econometric aspects of financial time series and different ways of modeling them are presented, e.g., in Gourieroux and Jasiak (2001).

changes in long- and short-term interest rates, inflation, the real growth rate of the economy, and market sentiment."[4]

Macroeconomic factors are appealing because they capture the fundamentals of the general economic situation, they are likely to be stable over time, they are easily available, and the "standard" selection of factors tends to be suitable for most stocks. However, macroeconomic factors usually come with two major downsides: (i) much like the CAPM's sole use of the market index, they focus on general market aspects and do not leave much room for industry or firm specifics, and (ii) factors such as inflation are not available on a short term basis, hence short term effects might be neglected and the estimation of the factor weights demands data over a long period of time assuming stationarity.

Other authors therefore use factors that are likely to be firm or industry specific.[5] Such tailor-made sets of factors can be expected to outperform a bundle based on rather general guidelines. As with the selection of macroeconomic factors, the bundles of firm specific factors are generally found in two ways: they are either selected *a priori*, based on economic or fundamental considerations, or they are found by some statistical optimization process. Either alternative has advantages and disadvantages: the former ensures quick results, but reliable factors might be excluded too early; the latter might produce models that allow for better estimations, yet the intuition behind the found combination might get lost.

Though not desirable from an economists' point of view, selecting factors solely on statistical grounds might be reasonable in several circumstances, e.g., when there is not enough information on the considered firm or data series (or, to be more precise, the cost of gathering and evaluating this information is too high); when there are not enough fundamental data to match and find reliable factor weights; when short term dependencies outweigh long term relationships; or when the number of equally plausible (or implausible) factors has to be reduced to avoid overfitting. In addition, statistical factor selection can help to get a first idea of which factors might or might not have a link to some asset's prices. However, it does not guarantee to identify fundamental and economically plausible relationships.

---

[4] See www.birr.com and www.birr.com/sector.cfm.

[5] See., e.g., the results in King (1966) (that actually precede the APT) or, more recently, Berry, Burmeister, and McElroy (1988), Fama and French (1992) and Brennan, Chordia, and Subrahmanyam (1998).

Model selection is a computationally demanding task. Not least due to the large number of alternatives that usually comes with the factor selection problem, statistical approaches tend to start with a rather small pool of available factors which has undergone an *a priori* selection. Pesaran and Timmermann (1995), e.g., restrict themselves to nine factors from which they can select any combination of regressors and find the optimum by testing all possible solutions. For larger problems, however, complete enumeration is not possible and alternatives have to be found. Winker (2001), e.g., shows that optimization heuristics are one way to come to grips with model selection problems.

Based on Maringer (2004),[6] the main goal of this chapter is to present a method that allows for searching relevant risk factors without *a priori* knowledge of the firm's activities and without *a priori* reductions of the problem space. Section 7.2 formalizes the problem and describes the heuristic optimization approach that will be employed. The results of a computational study are reported in section 7.3. The chapter concludes with a critical discussion of the results and offers an outlook on possible extensions.

7.2   The Selection of Suitable Factors

7.2.1   The Optimization Problem

Let r_{ft} and r_{at} denote the return of factor f and the asset a, respectively, in period t. Given the validity of the APT and the underlying concepts,[7] asset a's return can be expressed according to

$$r_{at} = b_{a0} + \sum_{f \in B} b_{af} \cdot r_{ft} + u_{at}$$

---

[6] This chapter is a revised version of Maringer, D. (2004b), "Finding Relevant Risk Factors in Asset Pricing," *Computational Statistics and Data Analysis*, 47(2), 339–352. Elsevier's permission to reprint is gratefully acknowledged.

[7] Strictly speaking, this process and the assumed validity of a LRM are the underlying concepts in Ross's derivation rather than the result. The relationship between a LRM and the APT is discussed in more in detail, e.g., in Sharpe, Alexander, and Bailey (2003).

where b_{af} is the estimated sensitivity of r_{at} to the return of factor f and comes from a standard OLS multiple regression. b_{a0} is the estimated intercept, here introduced mainly for statistical reasons.[8] u_{at} is the residual with $u_{at} \sim N\left(0, \sigma_{u_a}^2\right)$.

The goal of the optimization problem is to find a bundle $\mathcal{B}$ that consists of k factors and that explains as much of r_a as possible. For the computational study, we used factors some of which have a close to perfect pair-wise correlation. Having such a couple in the bundle, the regression (if possible) might yield factor weights that are far from intuitively plausible, and the models lose reliability when out of sample tests are performed. As a simple measure to avoid these undesired effects of multi-collinearity, an upper limit, ρ_{max}^2, to the squared correlation between any two selected factors f and g in the bundle, ρ_{fg}^2, is introduced.

The quality of the model is measured by R^2.

The optimization problem for asset a can be stated as follows:

$$\max_{\mathcal{B}} R_a^2 = 1 - \frac{\Sigma_t u_{at}^2}{\Sigma_t \left(r_{at} - \bar{r}_a\right)^2}$$

subject to

$$u_{at} = r_{at} - \left(b_{a0} + \sum_{f \in \mathcal{B}} b_{af} \cdot r_{ft}\right)$$

$$|\mathcal{B}| = k$$

$$b_a = [b_{a0}\ b_{a1}\ \ldots\ b_{ak}]' = \left(f'f\right)^{-1} f'a$$

with $f = \begin{bmatrix}1 & [r_{ft}]_{f \in \mathcal{B}}\end{bmatrix}$ and $a = [r_{at}]$

$$\rho_{fg}^2 \leq \rho_{\text{max}}^2 \quad \forall f, g \in \mathcal{B} \text{ and } f \neq g$$

where $\bar{r}_a$ is the arithmetic mean return of asset a. u_{at} is the residual, i.e., the part of the asset's return not captured by the regression model. f and a are the matrix and vector of factor and asset returns, respectively; b_a are the factor weights for asset a subject to the selected factors and coming from a standard linear regression model.

---

[8] In passing, note that given an equilibrium situation, b_{a0} also captures the risk-free return; see equations (1.22) and (1.23).

7.2.2 Memetic Algorithms

Introduced by Pablo Moscato, *Memetic Algorithms* (*MA*) are a new meta-heuristic that combines local search with global search.[9] The basic idea is to have a population of cooperating, competing, and locally searching agents. The original idea is based on what Richard Dawkins calls *memes*, units of culture or knowledge that are self-replicating and changing with a tendency to spread.[10]

The local search part is based on the Simulated Annealing (SA) principle presented in Kirkpatrick, Gelatt, and Vecchi (1983). This method starts with some random structure and repeatedly suggests slight modifications. Modifications for the better are always accepted; modifications that lower the fitness of the objective function are accepted with a decreasing probability (depending on the magnitude of the impairment and the progress of the algorithm).

Unlike in other evolution based algorithms, MA does not rank its individuals according to their (relative) fitness, but arranges them in a fixed structure. Global search then includes competition where immediate neighbors challenge each other. Based on the same acceptance criterion as in the local search part, the challenger will impose her solution on the challenged agent. Global search also includes cooperation by a standard cross-over operation with mating couples chosen due to their position within the population's structure.

Applied to the factor selection problem, each individual of the population starts off with a feasible bundle of factors. *Local search* means that one or two of the current bundle's factors are exchanged heuristically for other factors (without violating the constraints).[11]

As indicated, agents *cooperate* by mating with some other agent which (unlike in most evolutionary algorithms) is not chosen according to her fitness but to her

---

[9] See Moscato (1989, 1999) and the presentation in section 2.3.4.

[10] See Dawkins (1976).

[11] For the case that k is an upper limit rather than a fixed value, the algorithm does not necessarily exchange factors but excludes some and/or includes others such that the constraint $|\mathcal{B}| \leq k$ is not violated. In this case, an alternative objective function to be maximized would be the adjusted goodness of fit measure, $R_{adj}^2 = 1 - (1 - R^2) \cdot (N - 1)/(N - |\mathcal{B}|)$ where N is the number of in sample observations.

"geographic" position. If the population consists of P agents (with P being an even number and each agent having a fixed index), the agent $i = 1 \ldots P/2$ will mate with agent $j = i + P/2$. The two children are generated by a cross-over operator[12] where each offspring inherits one part of the mothering and the other of the fathering agent (given that none of the constraints is violated). *Competition* resembles the "survival of the fittest" concept as any agent i is challenged by her neighbor $n = i + 1$. With the acceptance criterion from the local search part, i has a chance of withstanding n (i.e., i will not accept n's solution) only if n's fitness is lower than i's.

The algorithm was implemented using Matlab 6 on a Pentium IV 1.8 GHz. Each problem was solved repeatedly and independently, the reported results refer to the best solution of these runs. The computational complexity of the algorithm is given by the estimation of the model weights for a given selection and therefore quadratic in k. The CPU time needed for an independent optimization run was about 20 seconds for $k = 5$.

7.3   Computational Study

7.3.1   Data and Model Settings

In order to apply the methodology to the selection problem as presented in the previous section, we use daily data of the S&P 100 stocks over the horizon of November 1995 through November 2000. For the main study, these data were split into six time frames consisting of 200 trading days for the (in sample) regression plus 100 trading days for an (out of sample) test period. Since all the models are estimated separately, we allow the out of sample period to overlap with the next problem's in sample period. The set of candidate factors is made up of 103 Morgan Stanley Capital International (MSCI) Indices. These factors include 23 country indices, 42 regional indices,[13] and 38 industry or sector indices. The parameters for each period were selected independently from results for other periods. Due to incomplete data, two

---

[12]   See, e.g., Fogel (2001).

[13]   As indicated earlier, some of the regional indices are rather similar to each other and therefore have a high correlation.

problems had to be excluded, resulting in 598 problems to be solved. The upper limit for the pair-wise correlation for any pair of factors included is set to $\rho^2_{max} = 0.5$.

In their review of the literature Lehmann and Modest (1987) find that the number of factors has a rather small influence on the model estimates. Increasing the number of factors included in the model decreases the residual risk of the estimates and ultimately shifts the model towards the seemingly ideal situation where any (past) movement in r_a can be attributed to some factor (or a set of factors).[14] A high number of included factors reflects the original APT idea of perfect replication in a complete market; yet it would also call for longer data series[15] to avoid the peril of overspecification: The resulting models might then well describe the past but cannot necessarily explain or predict equally reliably. From a practical point of view, the models become less easily applicable the more factors are included. In the main part of our analysis, we therefore limit the number of included factors to $k = 5$ which is also a common choice in the literature.

7.3.2 Main Results for the Selection of Factors

7.3.2.1 The Goodness of Fit

From a statistical point of view, the available factors do have explanatory power for the securities' returns. On average, about 48 per cent of past variance in the daily price changes could be explained with the applied LRM, and for about 80 per cent of the problems, a model with an R^2 of at least 0.3 could be found. Approximately six per cent of the reported models have an R^2 of 0.8 or even more. In more than 90 per cent of all reported models, at least four of the five factor weights are statistically significant,[16] with all factor weights significant being the "standard" case. No model with just one significant factor was reported, and in only five out of the 598 models factor bundles with just two significant factors were found to be optimal. The constant factor b_{0i}, on the other hand was significant in just 9 cases. Figure 7.1 summarizes the means and the bandwidths for the R^2's for all companies.

---

[14] BARRA, a major risk management company, include 13 risk factors and 55 industry groups in their second generation U.S. Equity model, known as "E2". See www.barra.com and Rosenberg (1984).

[15] See also Shukla (1997) and the references therein.

[16] In this chapter, the level of confidence for statistical significance is 5 per cent.

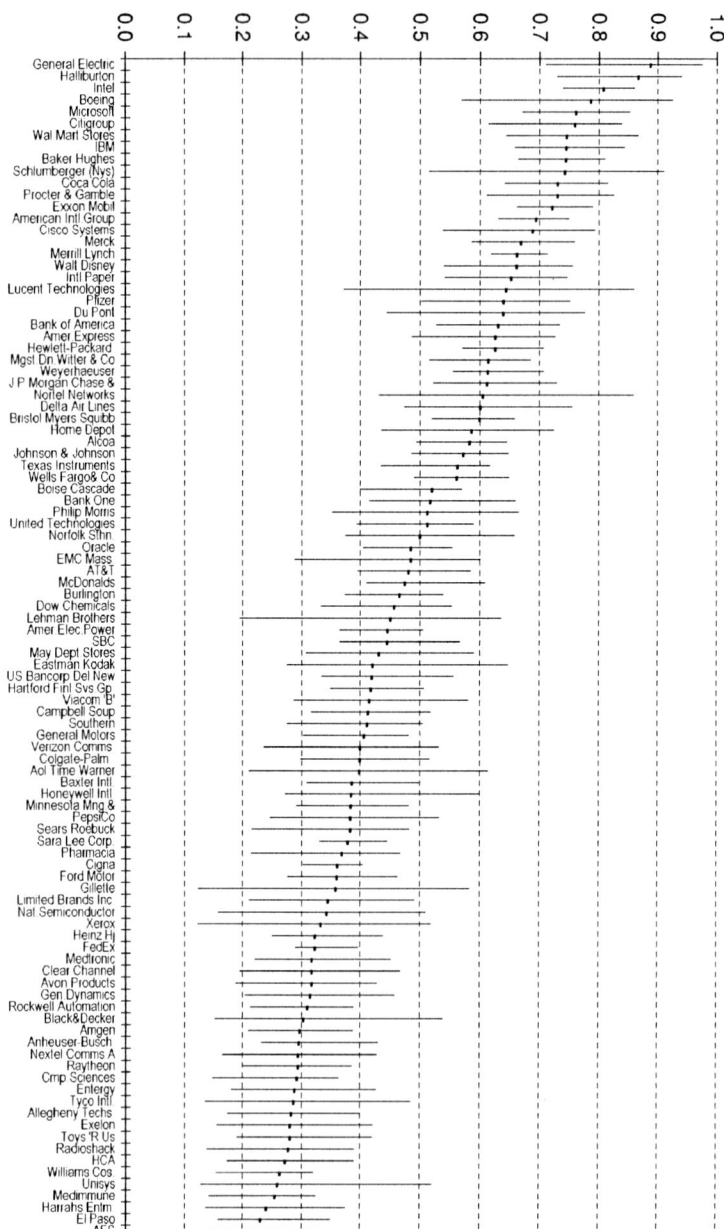

Fig. 7.1: Mean and bandwidth of companies' R^2's with $\rho^2_{\max} = 0.5$

7.3.2.2 Composition of the Optimal Bundles of Factors

Although the optimization was run separately for each asset and time frame, there were some general patterns that could be found in many solutions. One of these results is that sector indices are preferred over regional and country indices. The typical bundle consists of at least two, usually three sector indices, but just one or two regional and/or country indices. As industry and sector indices represent just about a third of the available indices (38 out of 103), the preference for this type of factors is not only highly significant yet also crucial to the quality of the results: In virtually all of the results, it is one of the sector indices that contributes most to the model's R^2. The selected industry factors tend to include at least one that represents the main area of activity of the considered firm, or one that is related to it, whereas the other industry factors might or might not have an apparent relationship to the company's operations. Figure 7.2 summarizes the frequencies at which the factors are chosen.

Countries appear slightly preferred over regional indices. However, neither of them are necessarily obvious choices: it comes rather surprising that "New Zealand" and "Finland" are the country factors chosen most often, followed by "Canada," "Venezuela" and "Brazil." The same applies for some of the regional indices, though to a smaller degree. Here, the respective factors for "The Americas," "North America," and "Europe ex EMU" are the ones chosen most often and that tend to contribute more than country indices. In this respect, the inclusion of certain countries might be regarded as a (more or less plausible) correction term (or a proxy for such a term which may not be included due to the correlation constraint) – which might also explain the sometimes unexpected signs in the parameters. The factor chosen least often is "The World Index" (though it is not the factor least correlated with the stock returns).

Typically, the bundles are not slimply a combination of the factors that would make the best single regressors: Ranking the factors according to their correlation with the stock, many a model uses just one or two factors with "top" correlation, whereas the others are "average" and usually at least one factor from those with least correlation (the latter usually being a geographical index). This is not necessarily a direct consequence of the constraint on the pair-wise correlation between included factors as alternative calculations with more or less generous values for $\rho_{\max}^2$ indicate.

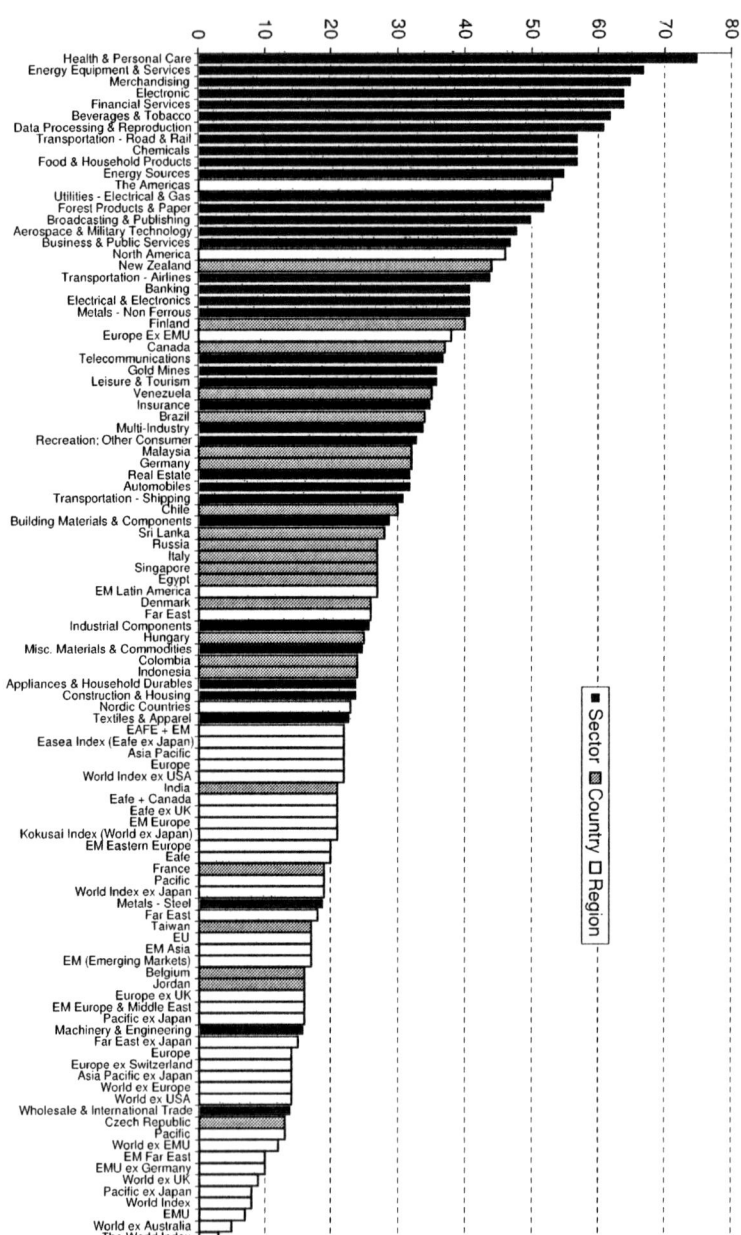

Fig. 7.2: Number of models a sector is included in

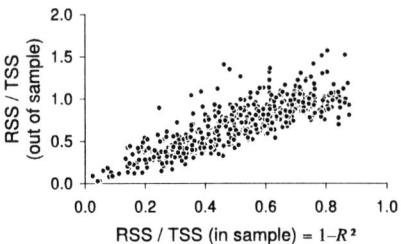

Fig. 7.3: Residual in sample and out of sample risk

7.3.2.3   Stability of the Results

Whether the chosen bundles for a given security remain the same over time does not necessarily depend on the goodness of fit for previous periods. In addition, the "plausibility" of bundles does not vary a lot over time. Typically, the dominant factors remain unchanged or are exchanged for other related sector indices (e.g., "Financial Services" might be exchanged for "Banking"). Fluctuation is found mostly in factors with low contribution to the model and in factors that owe their inclusion merely to spurious correlation or other statistical reasons. As a consequence, eyeballing might suggest that bundles change considerably from one period to another. The effects of these exchanges are notable, yet not as big as expected. Hence, keeping a bundle based on the in sample selection and performing an out of sample test with it shows that good explanatory power can be transformed into reasonably good predictions with low idiosyncratic risk (see Figure 7.3). As presented in section 4.3.1, the reliability of the results could be further improved by using more sophisticated estimates for the factor risks and returns; however, as this chapter is mainly concerned with the (in sample) selection problem, a more detailed discussion of these issues is left out.

Finding causes for stable or unstable R^2's appears to be a rather difficult task. We could not derive a general rule stating whether the models for a given company remain equally good over time. There appears to be no general pattern that indicates whether models with like factors or those with frequent changes in their bundles' composition show a more persistent R^2, nor could we isolate factors whose inclusion tends to indicate such a relationship. Though the models for the year 1998 appear to perform slightly better in some companies than those for other time windows, we could not find statistically significant evidence for temporal effects.

7.3.2.4 Number of Selected Factors

According to Trzcinka (1986), the number of significant factors increases with increasing sample size, but that the first factor remains dominant. This agrees with our finding: Typically, this "first" factor is a sector index as described earlier. Models where the dominant factor is geographical, on the other hand, tend to have less than average sector factors in their bundles – and to have a below average R^2. Also, some models include groups of indices that as a bundle increase a model's R^2 though individually show low relation to the security's return; yet these bundles rarely explain the lion's share of the model's goodness of fit.

The decreasing contribution of additional factors is also supported by further computational studies for our data: alongside the reported results for $k = 5$, we also ran models for any $k \leq 10$ and found that the factor that is selected for a single regression model is almost always found in the solutions for larger k as well, whereas second or third regressors might be replaced with a group of alternative factors rather than supplemented by simply adding an endorsing index.

Comparing the results for models with different k via the adjusted goodness of fit, $R^2_{adj} = 1 - (1 - R^2) \cdot (N - 1)/(N - |\mathcal{B}|)$ where N is the number of in sample trading days, shows that increasing the number of included factors is always advantageous – yet the marginal contribution of an additional factor is rapidly decreasing.[17] The results confirm that $k = 5$ is a reasonable choice: allowing for more than five factors brought hardly any substantial increase in R^2_{adj}. Quite often, this is already true for models with $k = 2$ or 3; in this case, however, the R^2_{adj} tends to be less stable over time. One reason might be that in these cases, the models include preferably factors that have a (more or less) fundamental relation to the asset but cannot capture temporal effects such as volatility clustering. Including factors with a less apparent

---

[17] An alternative to maximizing the R^2_{adj} would be the optimization of some information criterion combining residual risk and number of explanatory variables, see, e.g., Akaike (1974) or Schwarz (1978).

fundamental relationship (and low correlation with the asset) yet with supplementary characteristics might account for these aspects. These factors are more likely to be replaced over time than those with high correlation. Though temporal effects or their function as a proxy for unavailable factors can be one reason for the inclusion of these indices, spurious correlation might be another one. An interpretation of the economic contribution of these factors might therefore be misleading.

7.3.2.5   The Length of the In Sample Period

The results presented so far are for models where the in sample periods consist of 200 trading days. Increasing the period length has a limited effect on the results. As expected, the goodness of fit declines when the horizon length increases as there is less opportunity to adjust for temporal particularities (such as volatility clustering and structural breaks) – which might also be seen as a reduced peril of spurious correlation. In many cases the reduction in R^2 and R^2_{adj}, respectively, is therefore acceptable.

When the in sample number of days is 1 200, industry factors are still the preferred choice. Regions, however, are now more frequently chosen than with shorter time frames, and countries get less often included. In particular regional indices such as "North America," "The Americas" and "World Index ex USA" become popular which appears economically plausible given the considered assets. Although the optimization process does not (and, according to the problem statement, *cannot*) account for "economic plausibility" of the choices, the found solutions typically contain at least one factor that represents the company's industry – and one or two factors that show little correlation to the stock. For models with $k \leq 8$, "Finland" is still the most frequently chosen country index, and there are still models with factors where there is no apparent relationship or similarity between asset and chosen index. Yet again, these factors are usually included only when there are already "sensible" factors and they are not the main contributors to the models' goodness of fit. Generally, the qualitative results do not differ substantially from those for shorter time frames.

7.3.3   Alternative Models

The high correlation between some of the selected (industry) factors and the considered company might raise the question whether there is some endogeneity problem. Though no stock's weight in an index is large enough to cause correlation coefficients between factor and stock returns to be as high as 0.7 and even more, there are some companies that are typical representatives of their trade. Hence, they can have a high direct or indirect effect (via stocks which are highly correlated to them) on the factors which are then included in a model to explain the stock return.

We extend the original optimization problem and introduce a new constraint that allows only those factors f to enter a model that have a squared correlation with the return of stock a, ρ^2_{fa}, of 0.25 or less. As expected, the models with originally high R^2 see a sharp decline in the new model's goodness of fit: The factor(s) that correlate(s) best with the stock return's variability is (are) most likely to be excluded. The resulting models still show a clear preference for industry over geographic factors, but they lose explanatory power and, above all, plausibility. We therefore omit a detailed presentation of these results.

According to the initial considerations, the aim of this study is to find the optimal selection of indices when there is no prior knowledge about economical plausibility of the factors. The results indicate that the relevant industry factors are likely to be identified anyhow, but that they tend to come together with less plausible factors; an economic interpretation of all the results is therefore not always possible. If there actually is prior knowledge, such results can be avoided by obvious measures such as reducing the set of candidate factors to choose from or a low value for k as factors with low plausibility seem to be included as a supplement to plausible factors. Another way would be to extend the objective function by adding a punishment term that reflects the "degree of (im-)plausibility" and therefore accounts for a tradeoff between maximizing a statistical goodness of fit measure and explanatory power.

7.4   Conclusions

The main results from this chapter are that the factor selection problem associated with the APT can be approached with heuristic optimization, and that the returns of

S&P 100 stocks can be partially replicated with a small set of MSCI factors. Although we assumed that there is no *a priori* knowledge which factors ought to be included or are economically plausible, it turned out that the variability in the returns is often traced back to changes in either the company's or a related industry sector. Regional or country indices, on the other hand, have rather low explanatory power: when there is no suitable sector index and the model has to rely on geographical indices only, then the model will not perform very well.

As the factors used in the models were selected on statistical grounds only with the available set of factors differing from "standard" sets of factors used in the APT, some aspects remain to be discussed. The model used in this chapter is meant to be flexible enough to work without prior knowledge about the firms' fundamentals. Nonetheless, many of the found solutions contain dominant factors that are also economically plausible. At the same time, otherwise popular factors such as the World Index which are considered to be well diversified (and therefore ought to capture merely systematic risk) show up in fewer models than expected.

When the objective is the identification of fundamental economic factors, a lower statistical fitness (i.e., a lower R^2) might be acceptable when it comes with higher economic plausibility. For the chosen implementation, this modified optimization problem could be approached by adding a term to the objective function that "rewards" or "punishes" plausible or implausible bundles. "Implausible" factors (if not eliminated from the set of available factors in the first place) would then be accepted only when there is strong statistical support. However, this would require prior knowledge about the companies or considered time series.

Another question arising from our results is whether one can derive some general "rule of the thumb" for a good first guess for selecting factors. Though there seem to be some vague general patterns (such as the inclusion of industry factors), there is no "one bundle fits all" solution as the literature suggests for the use of macroeconomic factors (both over time and between companies) – but then, the individual models achieve higher R^2's than the models presented in the literature. Tests could therefore investigate whether there exists some "universal" set of MSCI indices or other factors that works reliable with most of the problems and in due course would make a better first guess than do the models based on macroeconomic factors. In particular, the forecasting qualities of the resulting models ought to be considered; though the results from section 4.3.1 (where the bundles identified in this chapter are combined with GARCH predictions for the factor returns) are quite promising, macroeconomic factors are sometimes claimed to be better predictable than market indices which might diminish the advantage of the MSCI based models.

For our computational study, the underlying set of factors was meant to capture either whole economies (regional indices) or the effects within certain industries (sector indices). These factors should be easily accessible and available at any point of time. As a consequence, an investor can estimate whether an asset will be affected by current (or expected) risks in one of these factors – and can therefore take adequate measures of risk management and for hedging. Nonetheless, the set of available factors might well be enhanced. With macroeconomic factors being quite popular in the literature as well as in real life applications, these could bring additional explanatory power. Different sorts of interest rates and exchange rates might be such data that are equally available as are the MSCI indices. Prices of goods and raw materials, on the other hand, might extend the list of factors that help explain firm specifics. The caveat, of course, is that providing too many factors to choose from increases the danger of data-fitting and selecting implausible bundles that are tailor-made for past data, but are unable to predict future returns.

All of these points might lead to further interesting and helpful insights about the application of the APT. The results in this chapter suggest that the chosen methodology might help in answering these questions.

Chapter 8

Concluding Remarks

The main objective of this book is to demonstrate how to use heuristic optimization techniques to approach portfolio optimization problems that could not be answered with traditional methods. With these new tools, it was possible to investigate the effects of market frictions which are usually ignored or appear in a rather stylized:

- In chapter 3, it is shown that transaction costs might lead to optimal portfolio weights that differ substantially from those under perfect market conditions. This is all the more true when another friction is included and only whole-numbered quantities of stocks can be purchased and the investors have different initial endowments. It is shown that ignoring these frictions might result in noticeably inferior solutions; the traditional way of first solving the problem without the frictions and then finding a "similar" solution that satisfies the constraints can even result in portfolios negative expected returns. On the other hand, considering these frictions in the optimization process can reduce the disprofits of transaction costs and limited initial endowment.

- Chapter 4 focuses on the diversification in small portfolios. Both the finance literature and institutional investors often state that most of the risk diversification can be done with a rather small number of different assets. Also, empirical findings indicate that investors seem to prefer portfolios with a rather few different assets. In the lack of suitable methods, however, the literature so far could offer just rough estimates for the marginal contribution of adding another asset to the portfolio and how to find the optimal small portfolio. The findings in this chapter support the view that a portfolio with a small number of different assets can be almost as well diversified as large portfolios without such cardinality constraints – provided the portfolio has undergone a reliable selection process.

- The computational complexity of portfolio optimization problems might quickly get out of hand when the "standard" optimization problems are enhanced with market frictions or specific constraints. Though general-purpose meta-heuristics are often able to solve problems that are unapproachable to traditional optimization techniques, it might be helpful to take the particularities of these optimization problems into account and develop a tailor-made heuristic search method. Chapter 5 demonstrates how a hybrid search methods can be designed that is able to solve the given problem highly reliably and efficiently.

- The finance literature as well as the investment industry often complain of the deficiencies of the assumption that asset returns are normally distributed. Furthermore, the reliability of associated "bandwidth" risk measures such as the volatility has been doubted. During the last decade, the "point" or quantile risk measure Value at Risk (VaR) has become a new standard not only because of its intuitive appeal, but also because it does not assume a particular distribution of the assets returns. The results in chapter 6, however, raise serious doubts whether this measure, in conjunction with empirical distributions, is a suitable alternative to the volatility concept even when the already known shortcomings of VaR do not apply: The VaR under empirical distribution can be reliably estimated only when not considered in the portfolio selection process. Having the VaR as an explicit risk constraint might have a damaging effect on the reliability of the risk estimations, in particular under empirical distributions, and the use of the volatility would be superior even when neither the assets' nor the resulting portfolios' returns are normally distributed.

- The finance literature offers a series of equilibrium and prediction models among which the Arbitrage Pricing Theory (APT) has gained considerable popularity in theory as well as practical application. According to the APT, the return process of an asset can be described as a linear combination of factor returns. As there is no general rule for how to find the optimal set of factors for a particular asset, some models use a large number of different factors – which is not desirable because of the peril of over-specification or misspecification due to the inclusion of actually irrelevant factors. Models with a reduced set of factors are restricted to either choose from a small preselection of potential candidates (because of the computational complexity of the associated model selection problem) or using a general bundle of factors. Chapter 7 shows that the model selection problem, too, can be answered with heuristic optimization.

In the main part of this contribution, well-known problems in financial management are addressed that cannot be answered satisfactorily with traditional approaches. Therefore, the finance literature so far had to make simplifying assumptions in order to keep the resulting optimization problems manageable. In this contribution, however, a different approach was chosen: rather than formulating and simplifying the models in a way to make them solvable with traditional solution methods, the models were stated to be as close to reality as possible while adopting new solution techniques to overcome the limits of traditional approaches.

In all of the empirical studies in this contribution, some of the basic models' aspects were still chosen to be rather simple in order to be able to isolate the effects of the considered market frictions. Hence, it was assumed in all portfolio selection problems, that the investor faces a single period horizon, decides rationally, has a precise idea of expected returns and risks, and so on. Nonetheless, it could be shown that even in these rather simple settings traditional approaches are limited in their reliability since simplifying assumptions on allegedly irrelevant aspects might lead to converse conclusions and clearly inferior (or even opposing) results.

Introducing new methods rather than simplifying assumptions, the shortcomings of the traditional models could be identified and new aspects could be investigated. At the same time it could be demonstrated which of the analyzed simplifications in financial optimization models are tolerable and which lead to severely wrong decisions when transferred into practical applications.

Starting with applications to problems merely from areas such as operations research and econometrics, meta-heuristics and related methods from soft computing and computational intelligence are attracting increasing attention in the economics and business literature which lead to a series of successful applications of these methods. In the finance literature, first applications covered aspects such as estimation and prediction, pricing, and simulation. As many of these computational approaches have an inherent optimization aspect, they can also be applied to financial optimization problems (such as the portfolio selection problems presented in this contribution). In addition, a number of financial (and, in general, economic) problems can be regarded optimization problems (such as the estimation problem presented in section 2.4 or the factor selection in chapter 7) and can therefore also be approached.

The results from the empirical studies presented in this book provided not only new insights to the discussed problems. They also demonstrate that meta-heuristics are capable of answering computationally demanding problems reliably and efficiently. Hence, the numerical studies could avoid simplifying assumptions that are necessary when traditional methods are used at the cost of distorting the result. The results in this contribution are therefore also promising as the applied methods are flexible in their application and can therefore easily be adopted for new problems that are beyond the limits of traditional analysis.

Bibliography

Aarts, E. H. L., and J. K. Lenstra (2003): "Introduction," in *Local Search in Combinatorial Optimization*, ed. by E. H. L. Aarts, and J. K. Lenstra, chap. 1, pp. 1–17. John Wiley & Sons Ltd., Chichester, 2nd edn.

Aarts, E. H. L., and P. J. M. van Laarhoven (1985): "Statistical Cooling: A General Approach to Combinatorial Optimization Problems," *Philips Journal of Research*, 40, 193–226.

Admati, A. R. (1985): "A Noisy Rational Expectations Equilibrium for Multi-Asset Securities Markets," *Econometrica*, 53(3), 629–657.

Akaike, H. (1974): "A New Look at the Statistical Model Identification," *IEEE Transactions on Automatic Control*, AC-19(6), 716–723.

Alexander, G. J., and A. M. Baptista (2001): "A VaR-Constrained Mean-Variance Model: Implications for Portfolio Selection and the Basle Capital Accord," Working paper, University of Minnesota and University of Arizona.

Alexander, G. J., and J. C. Francis (1986): *Portfolio Analysis*. Prentice-Hall, Englewood Cliffs, NJ, 3rd edn.

Althöfer, I., and K.-U. Koschnik (1991): "On the Convergence of Threshold Accepting," *Applied Mathematics and Optimization*, 24, 183–195.

Andersen, T. G., L. Benzoni, and J. Lund (2002): "An Empirical Investigation of Continuous-Time Equity Return Models," *The Journal of Finance*, 57(3), 1239–1284.

Andersson, F., H. Mausser, D. Rosen, and S. Uryasev (2001): "Credit Risk Optimization with Conditional Value-at-Risk Criterion," *Math. Programming*, Ser. B 89, 273–291.

Artzner, P., F. Delbaen, J.-M. Eber, and D. Heath (1999): "Coherent Measures of Risk," *Mathematical Finance*, 9(3), 203–228.

Arzac, E., and V. Bawa (1977): "Portfolio Choice and Equilibrium in Capital Markets with Safety-First investors," *Journal of Financial Economics*, 4(3), 277–288.

Ausiello, G., and M. Protasi (1995): "Local Search, Reducibility and Approximability of NP–Optimization Problems," *Information Processing Letters*, 54(2), 73–79.

Azoff, E. M. (1994): *Neural Network Time Series Forecasting of Financial Markets*. Wiley, Chichester, New York.

Barber, B. M., and T. Odean (2000): "Trading Is Hazardous to Your Wealth: The Common Stock Investment Performance of Individual Investors," *The Journal of Finance*, 55(2), 773–806.

———— (2003): "All that Glitters: The Effect of Attention and News on the Buying Behavior of Individual and Institutional Investors," Working paper, University of California, Davis and Berkeley.

Barr, R. S., B. L. Golden, J. P. Kelly, M. G. C. Resende, and W. R. Stewart, Jr. (1995): "Designing and Reporting on Computational Experiments with Heuristic Methods," *Journal of Heuristics*, 1, 9–32.

Basak, S., and A. Shapiro (2001): "Value-at-Risk-Based Risk Management: Optimal Policies and Asset Prices," *The Review of Financial Studies*, 14(2), 371–405.

Basel Committee on Banking Supervision (2003): "Consultative Document: The New Basel Capital Accord, Bank for International Settlement," available from: www.bis.org.

Berkowitz, J., and J. O'Brien (2002): "How Accurate Are Value-at-Risk Models at Commercial Banks?," *The Journal of Finance*, 57(3), 1093–1111.

Berry, M. A., E. Burmeister, and M. B. McElroy (1988): "Sorting Out Risks Using Known APT Factors," *Financial Analysts Journal*, 44(1), 29–42.

Best, M. J., and R. R. Grauer (1985): "Capital Asset Pricing Compatible with Observed Market Value Weights," *The Journal of Finance*, 40(1), 85–103.

———— (1991): "On the Sensitivity of Mean-Variance-Efficient Portfolios to Changes in Asset Means: Some Analytical and Computational Results," *The Review of Financial Studies*, 4(2), 315–342.

———— (1992): "Positively Weighted Minimum-Variance Portfolios and the Structure of Asset Expected Returns," *The Journal of Financial and Quantitative Analysis*, 27(4), 513–537.

Bezdek, J. C. (1992): "On the Relationship Between Neural Networks, Pattern Recognition and Intelligence," *International Journal of Approximate Reasoning*, 6, 85–107.

———— (1994): "What Is Computational Intelligence," in *Computational Intelligence: Imitating Life*, ed. by J. M. Zurada, R. K. Marks II, and C. J. Robinson, pp. 1–12. IEEE Press, Piscataway, NJ.

Bienstock, D. (1996): "Computational Study of a Family of Mixed-Integer Quadratic Programming Problem," *Mathematical Programming*, 74(2), 121–140.

Black, F. (1972): "Capital Market Equilibrium with Restricted Borrowing," *Journal of Business*, 45(3), 444–455.

———— (1986): "Noise," *The Journal of Finance*, 41(3), 529–543.

Black, F., M. C. Jensen, and M. Scholes (1972): "The Capital Asset Pricing Model: Some Empirical Tests," in *Studies of the Theory of Capital Markets*, ed. by M. C. Jensen. Praeger Publishers, Inc., New York.

Blackmore, S. (1999): *The Meme Machine*. Oxford University Press, Oxford.

Blume, M. E., and I. Friend (1975): "The Asset Structure of Individual Portfolios and Some Implications for Utility Functions," *The Journal of Finance*, 30(2), 585–603.

Bollerslev, T. (1986): "Generalized Autoregressive Conditional Heteroscedasticity," *Journal of Econometrics*, 31(3), 307–327.

Bollerslev, T., and E. Ghysels (1996): "Periodic Autoregressive Conditional Heteroscedasticity," *Journal of Business and Economic Statistics*, 14(2), 139–151.

Bonabeau, E., M. Dorigo, and G. Theraulaz (1999): *Swarm Intelligence. From Natural to Artificial Systems*. Oxford University Press, New York and Oxford.

Börsch-Supan, A., and A. Eymann (2000): "Household Portfolios in Germany," Working paper, University of Mannheim.

Brandimarte, P. (2002): *Numerical Methods in Finance. A MATLAB-Based Introduction*, Probability and Statistics. Wiley, New York.

Breeden, D. T. (1979): "An Intertemporal Asset Pricing Model with Stochastic Consumption and Investment Opportunities," *Journal of Financial Economics*, 7(3), 265–296.

Brennan, M., T. Chordia, and A. Subrahmanyam (1998): "Alternative Factor Specifications, Security Characteristics, and the Cross-Section of Expected Stock Returns," *Journal of Financial Economics*, 49(3), 345–373.

Brennan, M. J. (1970): "Taxes, Market Valuation and Corporate Financial Policy," *National Tax Journal*, 23(4), 417–427.

——— (1971): "Capital Market Equilibrium with Divergent Borrowing and Lending," *Journal of Financial and Quantitative Analysis*, 6(5), 1197–1205.

——— (1975): "The Optimal Number of Securities in a Risky Asset Portfolio when There Are Fixed Costs of Transacting: Theory and Some Empirical Results," *Journal of Financial and Quantitative Analysis*, 10(3), 483–496.

Brennan, M. J., and Y. Xia (2002): "Dynamic Asset Allocation under Inflation," *The Journal of Finance*, 57(3), 1201–1238.

Brock, W., J. Lakonishok, and B. LeBaron (1992): "Simple Technical Trading Rules and the Stochastic Properties of Stock Returns," *The Journal of Finance*, 47(5), 1731–1764.

Brooks, C. (2002): *Introductory Econometrics for Finance*. Cambridge University Press, Cambridge, UK.

Brooks, C., S. P. Burke, and G. Persand (2001): "Benchmarks and the Accuracy of GARCH Model Estimation," *International Journal of Forecasting*, 17(1), 45–56.

Buckley, I., G. Comezaña, B. Djerroud, and L. Seco (2003): "Portfolio Optimization when Asset Returns Have the Gaussian Mixture Distribution," Working paper, Center for Quantitative Finance, Imperial College, London.

Bullnheimer, B., R. Hartl, and C. Strauss (1999): "A New Rank Based Version of the Ant System — A Computational Study," *Central European Journal of Operations Research*, 7, 25–38.

Burmeister, E., and K. D. Wall (1986): "The Arbitrage Pricing Theory and Macroeconomic Factor Measures," *The Financial Review*, 21, 1–20.

Campbell, J. Y. (2000): "Asset Pricing at the Millennium," *The Journal of Finance*, 55(4), 1515–1567.

Campbell, J. Y., and A. S. Kyle (1993): "Smart Money, Noise Trading and Stock Price Behaviour," *The Review of Economic Studies*, 60(1), 1–34.

Campbell, J. Y., A. W. Lo, and A. C. MacKinlay (1997): *The Econometrics of Financial Markets*. Princeton University Press, Princeton, NJ, 2nd edn.

Campbell, J. Y., and T. Vuolteenaho (2003): "Bad Beta, Good Beta," Discussion Paper 2016, Harvard Institute of Economic Research, Cambridge, MA.

Campbell, R., R. Huisman, and K. Koedijk (2001): "Optimal Portfolio Selection in a Value-at-Risk Framework," *Journal of Banking and Finance*, 25(9), 1789–1804.

Cesari, R., and D. Cremonini (2003): "Benchmarking, Portfolio Insurance and Technical Analysis: A Monte Carlo Comparison of Dynamic Strategies of Asset Allocation," *Journal of Economic Dynamics & Control*, 27(6), 987–1011.

Chang, T.-J., N. Meade, J. Beasley, and Y. Sharaiha (2000): "Heuristics for Cardinality Constrained Portfolio Optimisation," *Computers and Operations Research*, 27(13), 1271–1302.

Chen, N.-F., R. R. Roll, and S. A. Ross (1986): "Economic Forces and the Stock Market," *Journal of Business*, 59(3), 383–404.

Cheng, P. L., and R. R. Grauer (1980): "An Alternative Test of the Capital Asset Pricing Model," *American Economic Review*, 70(4), 660–671.

Chiang, A. C. (1984): *Fundamental Methods of Mathemacial Economics*. McGraw-Hill, New York, London.

Colorni, A., M. Dorigo, and V. Maniezzo (1992a): "Distributed Optimization by Ant Colonies," in *Proceedings of the First European Conference on Artificial Life ECAL '91*, ed. by F. Varela, and P. Bourgine, pp. 134–142. Elsevier, Paris.

——— (1992b): "Investigation of Some Properties of an Ant Algorithm," in *Proceedings of the Parallel Problem Solving from Nature*, ed. by R. Männer, and B. Manderick, pp. 509–520. Elsevier, Brussels.

Corne, D., F. Glover, and M. Dorigo (eds.) (1999): *New Ideas in Optimization*. McGraw-Hill, London.

Crama, Y., and M. Schyns (2003): "Simulated Annealing for Complex Portfolio Selection Problems," *European Journal of Operational Research*, 150(3), 546–571.

Cvitanić, J., and I. Karatzas (1999): "On Dynamic Measures of Risk," *Finance and Stochastics*, 3(4), 451–482.

Daníelsson, J. (2002): "The Emperor Has No Clothes: Limits to Risk Modelling," *Journal of Banking & Finance*, 26(7), 1273–1296.

Daníelsson, J., P. Embrechts, C. Goodhart, C. Keating, F. Muennich, O. Renault, and H. S. Shin (2001): "An Academic Response to Basel II," Special Paper 30, Financial Markets Group, London School of Economics and Politial Science.

Daníelsson, J., B. N. Jorgensen, C. G. de Vries, and X. Yang (2001): "Optimal Portfolio Allocation Under a Probabilistic Risk Constraint and the Incentives for Financial Innovation," Tinbergen Institute discussion paper 01-069/2, Erasmus Universiteit, Amsterdam.

Dawkins, R. (1976): *The Selfish Gene*. Oxford University Press, Oxford.

de Beus, P., M. Bressers, and T. de Graaf (2003): "Alternative Investments and Risk Measurement," Working paper, Ernst & Young Acturaries, Asset Risk Management, Utrecht, The Netherlands.

De Giorgi, E. (2002): "A Note on Portfolio Selection under Various Risk Measures," Working paper, Risklab, ETH Zürich.

de Santis, G., and B. Gérard (1997): "International Asset Pricing and Portfolio Diversification with Time-Varying Risk," *The Journal of Finance*, 52(5), 1881–1912.

Dittmar, R. F. (2002): "Nonlinear Pricing Kernels, Kurtosis Preference, and Evidence from the Cross Section of Equity Returns," *The Journal of Finance*, 57(1), 369–403.

Dorigo, M. (1992): "Optimization, Learning and Natural Algorithms," Dissertation, Politecnico di Milano, Milano.

Dorigo, M., and G. Di Caro (1999): "The Ant Colony Optimization Meta-Heuristic," in *New Ideas in Optimization*, ed. by D. Corne, M. Dorigo, and F. Glover. McGraw-Hill, London.

Dorigo, M., V. Maniezzo, and A. Colorni (1991): "Positive Feedback as a Search Strategy," Technical Report 91-016, Politecnico di Milano, Milano, Italy.

―――― (1996): "The Ant System: Optimization by a Colony of Cooperating Agents," *IEEE Transactions on Systems, Man, and Cybernetics - Part B*, 26(1), 29–42.

Dueck, G., and T. Scheuer (1990): "Threshold Accepting: A General Purpose Algorithm Appearing Superior to Simulated Annealing," *Journal of Computational Physics*, 90, 161–175.

Dueck, G., and P. Winker (1992): "New Concepts and Algorithms for Portfolio Choice," *Applied Stochastic Models and Data Analysis*, 8, 159–178.

Duffie, D. (2001): *Dynamic Asset Pricing Theory*. Princeton University Press, Princeton, NJ, and Oxford, 3rd edn.

Dybvig, P. H. (1983): "An Explicit Bound on Deviations from APT pricing in a Finite Economy," *Journal of Financial Economics*, 12(4), 483–496.

Dyl, E. A. (1975): "Negative Betas: The Attractions of Short Selling," *Journal of Portfolio Management*, 1(3), 74–76.

Elton, E. J., and M. J. Gruber (1984): "Non-Standard CAPMs and the Market Portfolio," *The Journal of Finance*, 39(3), 911–924.

Elton, E. J., M. J. Gruber, S. J. Brown, and W. N. Goetzmann (2003): *Modern Portfolio Theory and Investment Analysis*. Wiley & Sons, Inc., Hoboken, NJ, 6th edn.

Engle, R. (1982): "Autoregressive Conditional Heteroscedasticity with Estimates of the Variance of United Kingdom Inflation," *Econometrica*, 50(1), 987–1007.

Engle, R. F. (ed.) (1995): *ARCH: Selected Readings*. Oxford University Press, Oxford, UK.

Fabozzi, F. J., and J. C. Francis (1978): "Beta as a Random Coefficient," *Journal of Financial and Quantitative Analysis*, 13(1), 101–114.

Fama, E. F. (1970): "Efficient Capital Markets: A Review of Theory and Empirical Work," *The Journal of Finance*, 25(2), 383–417.

Fama, E. F., and K. R. French (1992): "The Cross-Section of Expected Stock Returns," *The Journal of Finance*, 47(2), 427–465.

————— (1996): "The CAPM is Wanted, Dead or Alive," *The Journal of Finance*, 51(5), 1947–1958.

Fama, E. F., and J. D. MacBeth (1973): "Risk, Return, and Equilibrium: Empirical Tests," *Journal of Political Economy*, 81(3), 607–636.

Farrell, Jr., J. (1997): *Portfolio Management: Theory and Application*. McGraw-Hill, New York et al., 2nd edn.

Fiorentini, G., G. Calzolari, and L. Panattoni (1996): "Analytic Derivatives and the Computation of GARCH Estimates," *Journal of Applied Econometrics*, 11(4), 399–417.

Fogel, D. B. (2001): *Evolutionary Computation: Toward a New Philosophy of Machine Intelligence*. IEEE Press, New York, NY, 2nd edn.

Fogel, D. L. (ed.) (1998): *Evolutionary Computation: The Fossil Record*. IEEE Press, Piscataway, NJ.

Francis, J. C. (1975): "Skewness and Investors' Decisions," *Journal of Financial and Quantitative Analysis*, 10(1), 163–174.

Frey, R., and A. J. McNeil (2002): "VaR and Expected Shortfall in Portfolios of Dependent Credit Risks: Conceptual and Practical Insights," *Journal of Banking & Finance*, 26(7), 1317–1334.

Friend, E., Y. Landskroner, and E. Losq (1976): "The Demand for Risky Assets and Uncertain Inflation," *The Journal of Finance*, 31(5), 1287–1297.

Friend, I., and R. Westerfield (1980): "Co-Skewness and Capital Asset Pricing," *The Journal of Finance*, 35(4), 897–913.

Gaivoronski, A. A., and F. Stella (2003): "On-line Portfolio Selection Using Stochastic Programming," *Journal of Economic Dynamics & Control*, 27(6), 1013–1043.

Garey, M., and D. Johnson (1979): *Computers and Intractability – A Guide to the Theory of Incompleteness.* H.W. Freeman and Company, San Francisco.

Gençay, R., F. Selçuk, and B. Whitcher (2003): "Systematic Risk and Timescales," *Quantitative Finance*, 3(2), 108–116.

Gibbons, M. R. (1982): "Multivariate Tests of Financial Models: A New Approach," *Journal of Financial Economics*, 10(1), 3–27.

Gilli, M., and E. Këllezi (2002): "A Global Optimization Heuristic for Portfolio Choice with VaR and Expected Shortfall," in *Computational Methods in Decision-making, Economics and Finance*, ed. by E. Kontoghiorghes, B. Rustem, and S. Siokos, pp. 165–181. Kluwer.

——— (2003): "An Application of Extreme Value Theory for Measuring Risk," Discussion paper, University of Geneva.

Gilster, Jr., J. E. (1983): "Capital Market Equilibrium with Divergent Investment Horizon Length Assumptions," *Journal of Financial and Quantitative Analysis*, 18(2), 257–268.

Goldberg, J., and R. von Nitzsch (2001): *Behavioral Finance.* Wiley, Chichester.

Gonedes, N. (1976): "Capital Market Equilibrium for a Class of Heterogeneous Expectations in a Two-Parameter World," *The Journal of Finance*, 31(1), 1–15.

Goss, S., S. Aron, J. Deneubourg, and J. Pasteels (1989): "Self-Organized Shortcuts in the Argentine ant," *Naturwissenschaften*, 76, 579–581.

Gourieroux, C., and J. Jasiak (eds.) (2001): *Financial econometrics: Problems, Models, and Methods.* Princeton University Press, Princeton et al.

Gratcheva, E. M., and J. E. Falk (2003): "Optimal Deviations from an Asset Allocation," *Computers & Operations Research*, 30(11), 1643–1659.

Green, R. C. (1986): "Positively Weighted Portfolios on the Minimum-Variance Frontier," *The Journal of Finance*, 41(5), 1051–1068.

Grinblatt, M., and S. Titman (1983): "Factor Pricing in a Finite Economy," *Journal of Financial Economics*, 12(4), 497–507.

——— (1985): "Approximate Factor Structures: Interpretations and Implications for Empirical Tests," *The Journal of Finance*, 40(5), 1367–1373.

Grossman, B., and W. F. Sharpe (1984): "Factors in Security Returns," Discussion paper, Center for the Study of Banking and Financial Markets, University of Washington.

Grossman, S. J., and J. E. Stiglitz (1976): "Information and Competitive Price Systems," *American Economic Review*, 66(2), 246–253.

——— (1980): "On the Impossibility of Informationally Efficient Markets," *American Economic Review*, 70(3), 393–408.

Guiso, L., T. Jappelli, and D. Terlizzese (1996): "Income Risk, Borrowing Constraints and Portfolio Choice," *American Economic Review*, 86(1), 158–172.

Halmström, B., and J. Tirole (2001): "LAPM: A Liquidity-Based Asset Pricet Model," *The Journal of Finance*, 56(5), 1837–1867.

Hansen, L. P., J. Heaton, and E. G. J. Luttmer (1995): "Econometric Evaluation of Asset Pricing Models," *The Review of Financial Studies*, 8(2), 237–274.

Harel, D. (1993): *Algorithmics: The Spirit of Computing*. Addison-Wesley, Harlow et al., 2nd edn.

Haugen, R. A. (2001): *Modern Investment Theory*. Prenctice Hall, London et al., 5th edn.

Hertz, A., and M. Widmer (2003): "Guidelines for the Use of Meta-Heuristics in Combinatorial Optimization," *European Journal of Operational Research*, 151(2), 247–252.

Hillier, F. S., and G. J. Lieberman (1995): *Introduction to Mathematical Programming*. McGraw-Hill, New York, London, 2nd edn.

——— (2003): *Introduction to Operations Research*. McGraw-Hill, New York, London, 7th edn.

Holland, J. H. (1975): *Adaption in Natural and Artificial Systems*. University of Michigan Press, Ann Arbor, MI.

Huang, C.-f., and R. H. Litzenberger (1988): *Foundations for Financial Economics*. North-Holland Publishing Company, New York, Amsterdam, London.

Huang, H.-H. (2004): "Comment on 'Optimal Portfolio Selection in a Value-at-Risk Framework'," *Journal of Banking & Finance*, forthcoming.

Hull, J. C. (2003): *Options, Futures, and Other Derivatives*. Prentice-Hall, Upper Saddle River, NJ, 5th edn.

Ingersoll, Jr., J. E. (1984): "Some Results in the Theory of Arbitrage Pricing," *The Journal of Finance*, 39(4), 1021–1039.

Jagannathan, R., and T. Ma (2003): "Risk Reduction in Large Portfolios: Why Imposing the Wrong Constraints Helps," *The Journal of Finance*, 58(4), 1651–1638.

Jansen, R., and R. van Dijk (2002): "Optimal Benchmark Tracking with Small Portfolios," *The Journal of Portfolio Management*, 28(2), 9–22.

Johnson, N. L. (1949): "Systems of Frequency Curves Generated by Methods of Translation," *Biometrika*, 36(1/2), 149–176.

Jorion, P. (2000): *Value at Risk*. McGraw-Hill, Chicago et al., 2nd edn.

Judd, K. L. (1998): *Numerical Methods in Economics*. MIT Press, Cambridge, MA.

Kahn, M. N. (1999): *Technical Analysis Plain & Simple: Charting the Markets in Your Language*. Financial Times/Prentice Hall, London.

Keber, C., and D. Maringer (2001): "On Genes, Insects and Crystals: Determing Marginal Diversification Effects with Nature Based Methods," presentation, Annual Meeting of the Society of Computational Economics at Yale University.

Keenan, D. C., and A. Snow (2002): "Greater Downside Risk Aversion," *The Journal of Risk and Uncertainty*, 24(3), 267–277.

Kellerer, H., U. Pferschy, and D. Pisinger (2004): *Knapsack Problems*. Springer, New York.

King, B. F. (1966): "Market and Industry Factors in Stock Price Behavior," *Journal of Business*, 39(1), 139–190.

Kirkpatrick, S., C. Gelatt, and M. Vecchi (1983): "Optimization by Simulated Annealing," *Science*, 220(4598), 671–680.

Kirman, A. P. (1992): "Whom or What Does the Representative Individual Represent?," *Journal of Economic Perspectives*, 6(2), 117–136.

Knuth, D. E. (1997): *The Art of Computer Programming*, vol. 1: Fundamental Algorithms; Vol. 2: Seminumerical Algorithms. Addison-Wesley, Reading, MA, Harlow, 3rd edn.

Kontoghiorghes, E., B. Rustem, and S. Siokos (eds.) (2002): *Computational Methods in Decision-making, Economics and Finance*. Kluwer.

Kothari, S., J. Shanken, and R. G. Sloan (1995): "Another Look at the Cross-Section of Expected Returns," *The Journal of Finance*, 50(1), 185–224.

Krokhmal, P., J. Palmquist, and S. Uryasev (2001): "Portfolio Optimization with Conditional Value-at-Risk Objective and Constraints," Working paper, University of Florida.

Kroll, Y., H. Levy, and A. Rapoport (1988): "Experimental Tests of the Separation Theorem and the Capital Asset Pricing Model," *The American Economic Review*, 78(3), 500–519.

Kunreuther, H., and M. Pauly (2004): "Neglecting Disaster: Why Don't People Insure Against Large Losses?," *The Journal of Risk and Uncertainty*, 28(1), 5–21.

Lawson, T. (1997): *Economics and Reality*, Economics as social theory. Routledge, London.

LeBaron, B. (1999): "Agent-Based Computational Finance: Suggested Reading and Early Research," *Journal of Economic Dynamics & Control*, 24(5–7), 679–702.

Lehmann, B., and D. Modest (1987): "The Economic Foundations of the Arbitrage Pricing Theory," *Journal of Financial Economics*, 21(2), 213–254.

Lettau, M., and H. Uhlig (1999): "Rules of Thumb versus Dynamic Programming," *American Economic Review*, 89(1), 148–174.

Levy, H. (1978): "Equilibrium in an Imperfect Market: A Constraint on the Number of Securities in the Portfolio," *The American Economic Review*, 68(4), 643–658.

———— (1997): "Risk and Return: An Experimental Analysis," *International Economic Review*, 38(1), 119–149.

Lewellen, J., and J. Shanken (2002): "Learning, Asset-Pricing Tests, and Market Efficiency," *The Journal of Finance*, 57(3), 1113–1145.

Liechty, M. W. (2003): "Covariance Matrices and Skewness: Modeling And Applications In Finance," PhD thesis, Institute of Statistics and Decision Sciences, Duke University, Durham, NC.

Lindenberg, N. (1979): "Capital Market Equilibrium with Price Affecting Institutional Investors," in *Portfolio Theory, 25 Years After*, ed. by E. J. Elton, and M. J. Gruber. North-Holland Publishing Company, Amsterdam.

Lintner, J. (1965): "The Valuation of Risk Assets and the Selection of Risky Investments in Stock Portfolios and Capital Budgets," *Review of Economics and Statistics*, 47(1), 13–37.

———— (1971): "The Effect of Short Selling and Margin Requirements in Perfect Capital Markets," *Journal of Financial and Quantitative Analysis*, 6(5), 1173–1195.

Lo, A. W., and A. C. MacKinlay (1990): "Data-Snooping Biases in Tests of Financial Asset Pricing Models," *The Review of Financial Studies*, 3(3), 431–467.

Lo, A. W., H. Mamaysky, and J. Wang (2000): "Foundations of Technical Analysis: Computational Algorithms, Statistical Inference, and Empirical Implementation," *The Journal of Finance*, 55(4), 1705–1765.

Loewenstein, M. (2000): "On Optimal Portfolio Trading Strategies for an Investor Facing Transactions Costs in a Continuous Trading Market," *Journal of Mathematical Economics*, 33(2), 209–228.

Longin, F. M. (2000): "From Value at Risk to Stress Testing: The Extreme Value Approach," *Journal of Banking & Finance*, 24(7), 1097–1130.

Lucas, A., and P. Klaasen (1998): "Extreme Returns, Downside Risk, and Optimal Asset Allocation," *The Journal of Portfolio Management*, 25(1), 71–79.

MacKinlay, A. C., and L. Pástor (2000): "Asset Pricing Models: Implications for Expected Returns and Portfolio Selection," *The Review of Financial Studies*, 13(4), 883–916.

Mandelbrot, B. B. (1997): *Fractals and Sacling in Finance. Discontinuity, Concentration, risk*, vol. E of *selecta*. Springer, New York.

Manganelli, S., and R. F. Engle (2001): "Value at Risk Models in Finance," Working paper, European Central Bank and New York University, March.

Maringer, D. (2001): "Optimizing Portfolios with Ant Systems," in *International ICSC Congress on Computational Intelligence: Methods and Applications (CIMA 2001)*, pp. 288–294. ICSC.

——— (2002a): "Portfolioselektion bei Transaktionskosten und Ganzzahligkeitsbeschränkungen," *Zeitschrift für Betriebswirtschaft*, 72(11), 1155–1176.

——— (2002b): "Wertpapierselektion mittels Ant Systems," *Zeitschrift für Betriebswirtschaft*, 72(12), 1221–1240.

——— (2003a): "Distribution Assumptions and Risk Constraints in Portfolio Optimization," Discussion Paper No. 2003-008E, Faculty of Economics, Law and Social Sciences, University of Erfurt.

——— (2003b): "Great Expectations and Broken Promises. Risk Constraints and Assumed Distributions in Portfolio Optimization," Limassol, Cyprus. Society for Computational Economics.

——— (2004): "Finding the Relevant Risk Factors for Asset Pricing," *Computational Statistics and Data Analysis*, 47, 339–352.

——— (2005): "Distribution Assumptions and Risk Constraints in Portfolio Optimization," *Computational Management Science*, forthcoming.

Maringer, D., and H. Kellerer (2003): "Optimization of Cardinality Constrained Portfolios with a Hybrid Local Search Algorithm," *OR Spectrum*, 25(4), 481–495.

Maringer, D., and P. Winker (2003): "Portfolio Optimization under Different Risk Constraints with Modified Memetic Algorithms," Discussion Paper No. 2003-005E, Faculty of Economics, Law and Social Sciences, University of Erfurt.

Markowitz, H. M. (1952): "Portfolio Selection," *The Journal of Finance*, 7(1), 77–91.

Markowitz, H. M., and A. F. Perold (1981): "Portfolio Analysis with Factors and Scenarios," *The Journal of Finance*, 36(14), 871–877.

Mayers, D. (1972): "Nonmarketable Assets and Capital Market Equilibrium under Uncertainty," in *Studies in Theory of Capital Markets*, ed. by M. C. Jensen. Praeger Publishers, Inc., New York.

——— (1973): "Nonmarketable Assets and the Determination of Capital Asset Prices in the Absence of a Riskless Asset," *Journal of Business*, 46(2), 258–267.

McCloskey, D. N. (1996): *The Vices of Economists – The Virtues of the Bourgeoisie*. Amsterdam University Press, Amsterdam.

——— (1998): *The Rethoric of Economics*. University of Wisconsin Press, Madison, Wis., 2nd edn.

McCullough, B. D., and C. G. Renfro (1999): "Benchmarks and Software Standards: A Case Study of GARCH procedures," *Journal of Economic and Social Measurement*, 25(2), 59–71.

Merton, R. (1973): "An Intertemporal Capital Asset Pricing Model," *Econometrica*, 41(5), 867–888.

Merton, R. C. (1969): "Lifetime Portfolio Selection under Uncertainty: The Continuous Time Case," *Review of Economics and Statistics*, 51(3), 247–257.

———— (1971): "Optimum Consumption and Portfolio Rules in a Continuous-Time Model," *Journal of Economic Theory*, 3(4), 373–413.

———— (1972): "An Analytic Derivation of the Efficient Portfolio Frontier," *Journal of Financial and Quantitative Analysis*, 7(4), 1851–1872.

———— (1992): *Continuous-Time Finance*. Blackwell, Cambridge, MA, rev. edn.

Michalewicz, Z., and D. B. Fogel (1999): *How to Solve It: Modern Heuristics*. Springer, New York.

Moscato, P. (1989): "On Evolution, Search, Optimization, Genetic Algorithms and Martial Arts: Towards Memetic Algorithms," Report 790, Caltech Concurrent Computation Program.

Moscato, P. (1999): "Memetic Algorithms: A Short Introduction," in *New Ideas in Optimization*, ed. by D. Corne, M. Dorigo, and F. Glover, pp. 219–234. MacGraw-Hill, London.

Mossin, J. (1966): "Equilibrium in a Capital Asset Market," *Econometrica*, 34(4), 768–783.

Nilsson, B., and A. Graflund (2001): "Dynamic Portfolio Selection: The Relevance of Switching Regimes and Investment Horizon," Working paper, Department of Economics, Lund University.

Oehler, A. (1995): *Die Erklärung des Verhaltens privater Anleger. Theoretischer Ansatz und empirische Analysen*, Betriebswirtschaftliche Abhandlungen. Schäffer-Poeschel, Stuttgart.

Osman, I. H., and J. P. Kelly (1996): "Meta-Heuristics: An Overview," in *Meta-Heuristics: Theory & Application*, ed. by I. H. Osman, and J. P. Kelly, chap. 1, pp. 1–21. Kluwer Academic Publisher, Boston/Dordrecht/London.

Osman, I. H., and G. Laporte (1996): "Metaheuristics: A Bibliography," *Annals of Operations Research*, 63, 513–623.

Pesaran, M. H., and A. Timmermann (1995): "Predictability of Stock Returns: Robustness and Economic Significance," *The Journal of Finance*, 50(4), 1201–1228.

Peters, E. E. (1996): *Chaos and Order in the Capital Markets: A New View of Cycles, Prices, and Market Volatility*. Wiley, New York et al., 2nd edn.

Pflug, G. (2000): "Some Remarks on the Value-at-Risk and the Conditional Value-at-Risk," in *Probabilistic Constrained Optimization: Methodology and Applications*, ed. by S. Uryasev, pp. 278–287. Kluwer.

Pogue, G. A. (1970): "An Extension of the Markowitz Portfolio Selection Model to Include Variable Transactions' Costs, Short Sales, Leverage Policies and Taxes," *The Journal of Finance*, 25(5), 1005–1027.

Pritsker, M. (1997): "Evaluating Value at Risk Methodologies: Accuracy versus Computational Time," *Journal of Financial Services Research*, 12(2/3), 201–242.

Rasmussen, N., P. S. Goldy, and P. O. Solli (2002): *Financial Business Intelligence: Trends, Technology, Software Selection, and Implementation*. Wiley, New York.

Rechenberg, I. (1965): "Cybernetic Solution Path of an Experimental Problem," Library Translation 1122, Royal Aircraft Establishment.

———— (1973): *Evolutionsstrategie: Optimierung technischer Systeme nach Prinzipien der biologischen Evolution*. Fromman-Holzboog Verlag, Stuttgart.

Richard, S. F. (1979): "A Generalized Capital Asset Pricing Model," in *Portfolio Theory, 25 Years After*, ed. by E. J. Elton, and M. J. Gruber. North-Holland Publishing Company, Amsterdam.

Riskmetrics Group (1996): *Riskmetrics(TM) – Technical Document*J. P. Morgan / Reuters, New York, 4th edn.

Rockafellar, R. T., and S. Uryasev (2000): "Optimization of Conditional Value at Risk," *Journal of Risk*, 2(3), 21–41.

———— (2002): "Conditional Value-at-Risk for General Loss Distributions," *Journal of Banking & Finance*, 26(7), 1443–1471.

Roll, R. R. (1977): "A Critique of the Asset Pricing Theory's Tests; Part I. On Past and Potential Testability of the Theory," *Journal of Financial Economics*, 4(2), 129–176.

———— (1978): "Ambiguity When Performance is Measured by the Securities Market Line," *The Journal of Finance*, 33(4), 1051–1069.

Roll, R. R., and S. A. Ross (1980): "An Empirical Investigation of the Arbitrage Pricing Theory," *The Journal of Finance*, 39(5), 1073–1104.

Rosenberg, B. (1984): "Prediction of Common Stock Investment Risk," *Journal of Portfolio Management*, 11(1), 44–53.

———— (1985): "Prediction of Common Stock Betas," *Journal of Portfolio Management*, 11(2), 5–14.

Rosenberg, B., and J. Guy (1976): "Prediction of Beta from Investment Fundamentals: Part II," *Financial Analysts Journal*, 32(4), 62–70.

Ross, S. A. (1976): "The Arbitrage Theory of Capital Asset Pricing," *The Journal of Economic Theory*, 13(3), 341–360.

———— (1977): "Return, Risk, and Arbitrage," in *Risk and Return in Finance*, ed. by I. Friend, and J. L. Bicksler, pp. 189–218. Ballinger, Cambridge, MA.

Roy, A. (1952): "Safety-First and the Holding of Assets," *Econometrica*, 20(3), 431–449.

Russell, S., and P. Norvig (2003): *Artificial Intelligence. A Modern Approach*. Prentice Hall, Englewood Cliffs, NJ, 2nd edn.

Schmid, F., and M. Trede (2003): "Simple Tests for Peakdness, Fat Tails and Leptokurtosis Based on Quantiles," *Computational Statistics & Data Analysis*, 43(1), 1–12.

Schwarz, G. (1978): "Estimating the Dimenson of a Model," *Annals of Statistics*, 6, 461–464.

Scott, R. C., and P. A. Horvath (1980): "On the Direction of Preference for Moments of Higher Order Than the Variance," *The Journal of Finance*, 35(4), 915–919.

Seydel, R. (2002): *Tools for Computational Finance*. Springer, Berlin et al.

Shanken, J. (1992): "On the Estimation of Beta-Pricing Models," *The Review of Financial Studies*, 5(1), 1–33.

Sharpe, W. F. (1963): "A Simplified Model for Portfolio Analysis," *Management Science*, 9(2), 277–293.

———— (1964): "Capital Asset Prices: A Theory of Market Equilibrium und Conditions of Risk," *The Journal of Finance*, 19(3), 425–442.

———— (1966): "Mutual Fund Performance," *Journal of Business*, 39(1), 119–138.

———— (1991): "Capital Asset Prices with and without Negative Holdings," *The Journal of Finance*, 46(2), 489–509.

———— (1994): "The Sharpe Ratio," *Journal of Portfolio Management*, 21(1), 49–58.

Sharpe, W. F., G. J. Alexander, and J. V. Bailey (2003): *Investments*. Prentice-Hall International, Upper Saddle River, NJ, 7th edn.

Shefrin, H. (ed.) (2001): *Behavioral Finance*, International library of critical writings in financial economics. Edward Elgar, Cheltenham.

Shiller, R. J. (ed.) (1989): *Market volatility*. MIT Press, Cambridge, MA.

———— (2000): *Market Exuberance*. Princeton University Press, Chichester.

Shukla, R. (1997): "An Empiricists Guide to the Arbitrage Pricing Theory," Working paper, Syracuse University, Syracuse, NY.

Silver, E. A. (2002): "An Overview of Heuristic Solution Methods," Working Paper 2002–15, Haskayne School of Business, University of Calgary.

Simons, K. (2000): "The Use of Value at Risk by Institutional Investors," *New England Economic Review*, 11/12, 21–30.

Solnik, B. (1973): "The Advantages of Domestic and International Diversification," in *International Capital Markets*, ed. by E. J. Elton, and M. J. Gruber. North-Holland Publishing Company, Amsterdam.

———— (1974): "Why Not Diversify Internationally Rather than Domestically?," *Financial Analysts Journal*, 30(4), 48–54.

Stambaugh, R. F. (1982): "On the Exclusion of Assets from Tests of the Two-Parameter Model: A Sensitivity Analysis," *Journal of Financial Economics*, 10(3), 237–268.

Stepan, A., and E. O. Fischer (2001): *Betriebswirtschaftliche Optimierung: Einführung in die quantitative Betriebswirtschaftslehre*. Oldenbourg, München, 7th edn.

Stracca, L. (2004): "Behavioral Finance and Asset Prices: Where Do We Stand?," *Journal of Economic Psychology*, 25(3), 373–405.

Sullivan, R., A. Timmermann, and H. White (1999): "Data-Snooping, Technical Trading Rule Performance, and the Bootstrap," *The Journal of Finance*, 54(5), 1647–1691.

Sundaresan, S. M. (2000): "Continuous-Time Methods in Finance: A Review and an Assessment," *The Journal of Finance*, 55(4), 1569–1622.

Szegö, G. (2002): "Measures of Risk," *Journal of Banking & Finance*, 26(7), 1253–1272.

Taillard, E. D., L. M. Gambardella, M. Gendreau, and J.-Y. Potvin (2001): "Adaptive Memory Programming: A Unified View of Metaheuristics," *European Journal of Operations Research*, 135(1), 1–16.

Taylor, S. J. (1986): "Forecasting and the Volatility of Currency Exchange Rates," *International Journal of Forecasting*, 3(1), 159–170.

Thaler, R. H. (ed.) (1993): *Advances in Behavioral Finance*. Russell Sage Foundation, New York, NY.

Tobin, J. (1958): "Liquidity Preference as Behavior Towards Risk," *Review of Economic Studies*, 26(1), 65–86.

——— (1965): "The Theory of Portfolio Selection," in *The Theory of Interest Rates*, ed. by F. Hahn, and F. Brechling. Macmillan & Co. Ltd., London.

Trzcinka, C. A. (1986): "On the Number of Factors in the Arbitrage Pricing Model," *The Journal of Finance*, 41(2), 347–368.

Tsiang, S. (1972): "The Rationale of the Mean-Standard Deviation Analysis, Skewness Preference, and the Demand for Money," *Journal of Financial and Quantitative Analysis*, 6(3), 354–371.

Uryasev, S. (2000): "Conditional Value-at-Risk: Optimization Algorithms and Applications," *Financial Engineering News*, 14.

van der Saar, N. L. (2004): "Behavioral Finance: How Matters Stand," *Journal of Economic Psychology*, 25(3), 425–444.

Varian, H. (1993): "A Portfolio of Nobel Laureates: Markowitz, Miller and Sharpe," *The Journal of Economic Perspectives*, 7(1), 159–169.

Weber, M., and C. Camerer (1992): "Ein Experiment zum Anlegerverhalten," *Zeitschrift für betriebswirtschaftliche Forschung*, 44(2), 131–148.

Winker, P. (2001): *Optimization Heuristics in Econometrics. Applications of Threshold Accepting.* John Wiley & Sons, ltd., Chichester et al.

Winker, P., and M. Gilli (2004): "Applications of Optimization Heuristics to Estimation and Modelling Problems," *Computational Statistics & Data Analysis*, 47, 211–223.

Winker, P., and D. Maringer (2003): "The Hidden Risks of Optimizing Bond Portfolios under VaR," Sydney, Australia. 16th Australasian Finance and Banking Conference.

Zadeh, L. A., and J. M. Garibaldi (eds.) (2004): *Applications and Science in Soft Computing*, Advances in Soft Computing. Springer, New York et al.

Index

Advances in Computational Management Science

Printed in the United Kingdom
by Lightning Source UK Ltd.
118197UK00001B/118